Irene

About the Author

MARY SOUTH was a founding editor of Riverhead Books and also worked at Houghton Mifflin; Ballantine; Little, Brown & Company; and Rodale. In the course of her career, she edited an eclectic list of award-winning and bestselling books, including *The South Beach Diet*. When South is not aboard the *Bossanova*, she lives in New York City.

THE
CURE FOR
ANYTHING
IS SALT WATER

THE
CURE FOR
ANYTHING
IS SALT WATER

How I Threw My Life Overboard and
Found Happiness at Sea

MARY SOUTH

HARPER

NEW YORK · LONDON · TORONTO · SYDNEY

HARPER

Frontispiece courtesy of the author.

A hardcover edition of this book was published in 2007 by HarperCollins Publishers.

HarperCollins books may be purchased for educational, business, or sales promotional use. For information please write: Special Markets Department, HarperCollins Publishers, 10 East 53rd Street, New York, NY 10022.

FIRST HARPER PAPERBACK PUBLISHED 2008.

Designed by Jaime Putorti

Library of Congress Cataloging-in-Publication Data is available upon request.

ISBN 978-0-06-074703-9 (pbk.)

08 09 10 11 12   ID/RRD   10 9 8 7 6 5 4 3 2 1

To Vic, for keeping me afloat in so many ways
and
to Karyn, for my happy ending

The cure for anything is salt water—
sweat, tears or the sea.

—ISAK DINESEN

THE

CURE FOR

ANYTHING

IS SALT WATER

CHAPTER ONE

It's never too late to be who you might have been.
—GEORGE ELIOT

Not long ago, I was probably a lot like you. I had a successful career, a pretty home, two dogs and a fairly
normal life.

All I kept were the dogs.

Then one day in October 2003, I quit my good job and put
my sweet little house on the market. I packed a duffel bag of
clothes and everything else I owned went into storage. Within
weeks I was the proud owner of an empty bank account and a
40-foot, 30-ton steel trawler that I had no idea how to run. I
enrolled in nine weeks of seamanship school, and two weeks
after my course ended, I pulled away from the dock on my
very first trip: a 1,500-mile journey through the Atlantic from
Florida to Maine.

My transformation from regular person to unhinged mariner started casually enough. Lured to Pennsylvania a few

years ago by one more step up the book publishing career ladder, I had accepted a job that was editorial, managerial and very dull. I was busy enough at the office but, after work, I didn't know what to do with myself. I cooked, took guitar lessons, went to the gym, drank manhattans, watched movies at home and read books and magazines. But still I faced an abundance of excruciatingly quiet free time. On business trips to the city, I'd stock up on magazines. At first, I read a predictable assortment for a girl in exile from the big city: the *New Yorker, New York, New York Review of Books.*

Okay, it wasn't *all* about New York. There was *House and Garden, Dwell, Utne Reader, Maisons Côté Ouest, Vogue, Gourmet.* I'd read just about anything—which is probably how an occasional *Yachting* started to find its way into my stockpiles. When I saw *Motorboating, Sail* and *Powerboating* at the local supermarket, peeking out from behind the overwhelming number of firearm and bride publications (a combination that captured the flavor of the area all too well), I thought "Why not?" Soon, I had completely given up on literature, current events, even home decor. I started subscriptions to *Passagemaker* and *Soundings,* full year-long commitments. From there, it was a scary slide down the slippery slope to more extreme, niche titles (*Professional Mariner Magazine, Workboat Magazine, American Tugboat Review*) that I just *had* to have. I was becoming a trawler junky and I wasn't sure why.

But let's backtrack for a moment. I'd better start by admitting I am an optimist—not just your run-of-the-mill, happy-face, Pollyanna-type. I'm Old School—an *extreme* optimist of the sort that went out of style around the time of Don Quixote.

And like most optimists who regularly suffer the crushing defeats of a world less wonderful than they had imagined, I'm sure I have developed some finely honed coping strategies. (Or denial issues, if you prefer to call the glass half empty—as I obviously do not.) For instance, although I had just arrived at a new job in rural Pennsylvania full of vim and vigor, the deeply repressed realist within me knew almost immediately that I had made a terrible mistake. But there was no way I could admit that—even to myself.

The vocal Optimist in me said: *Hey, this is pretty cool. They have an organic café at work and the food's really inexpensive.*

But the mute Realist in me knew: Almost all of the food, no matter what it was, tasted weirdly *the same*, which—let's face it—was not good. At any price.

The Optimist said: *Wow. It's so rural out here that you'd never know you were only 100 miles from New York City.*

The Realist knew: I did not want to live in a place where the Wednesday Bob Evan's special was All the Possum You Can Eat for $3.99.

The Optimist said: *What a gorgeous stone house I have found for a bargain price!*

The Realist knew: I was going to ruin the rustic exposed stone walls (and drastically lower the resale value) when I splattered my brains all over them after a slow decline into loneliness and alcoholism.

My point is, maybe I wasn't able to *admit* to myself that I wanted out of that place in the worst possible way but nothing could have been less appropriate to my rural, landlocked situation than a sudden obsession with the boating lifestyle.

So perhaps my newfound passion was just a strangled cry for help, issued from the lonely wilds of scenic nowhere.

Every day, I'd put on a suit and drive to the office. I'd organize my editors, read submissions, review manuscripts, return phone calls from agents, do some editing, write and rewrite copy. I seemed to spend an inordinate amount of time in "brainstorming meetings" where a group of us, pulled away from whatever we'd been working on by a prearranged *ding* on our Outlook calendars, sat in a windowless, fluorescent-lit meeting room and tried to come up with just the right title for a health book. (It had to be prescriptive, it had to hold out a promise to the reader, it had to have punch. Using numbers was good. Dangling a plan was ideal. Thirty-Day Plans were…well…we were on fire.)

Once a week, the staff gathered for editorial meetings to decide which manuscripts we should buy. The sales director would weigh in with her department's assessment on the latest submissions, and it was uncanny how often they seemed to vote with one mind: hers—which was, sadly, as wide as a stream in Death Valley. As long as the author was a celebrity or at least had a well-established marketing platform, there was a possibility we could buy the book. Of course, there were other hurdles to clear. We wouldn't want to take any risks: the topic had to be fresh but not too fresh. In other words, someone needed to have published a book on the same subject, and sold enough copies to prove there was an audience but not so many as to suggest a been-there, done-that readership. Exactly what this number was varied with how much pressure the sales director was under from above, how things were going at home, how long her morning commute had taken and whether Mercury was out of retrograde.

And she wasn't the worst of it. My boss was a micromanager with an imagination that was significantly smaller than the stick up her butt. She was the classic corporate type: she put in long, long hours in a clever sleight that substituted endless meetings and frequent memos for actual productivity. But that's why they paid her the big bucks.

Anyway, the point is that innovation, new ideas, anything provocative or controversial was pretty much out of the question. There was a lot of talk about thinking outside the box, but at heart ours was an organization that liked a flow chart, a win-win, a net-net, everybody on the same page. In other words, I felt I had little to bring to the table. Don't get me wrong: I was a team player. But we didn't appear to have a team, just a cheerleading squad for the worst benchwarmers in the league.

I knew it wasn't *really* as bad as it seemed. It was corporate life. Not thrilling but a necessary evil. However, I was finding it increasingly intolerable. I felt, dully, that my soul was being quietly asphyxiated. And when I was released to the relative freedom of my tiny stone house each evening, all I could do was pour myself another manhattan, fire up the big screen and wish I was *anywhere* else. I knew it wasn't just the job. It was everything.

To be fair, I had always been a little of what my brother Hamilton disapprovingly called a "thrill-seeker." Nothing major: I backpacked through Europe, joined the Peace Corps, tried skydiving, sold my car and bought a motorcycle. Maybe my middle-age life and career were merely making me wistful for a sense of freedom that had become buried alive in a routine of dwindling satisfaction. As my office walls slowly disappeared behind a fleet of fishing boat photos, nautical charts and boat brochures, my wishes became increasingly specific: I wanted to be *there*, on that boat, or that one, and wherever they were going made absolutely no difference.

As my first Thanksgiving in Pennsylvania approached, my family in New York began the annual rite of passive-aggressive maneuvering to determine where we would all gather for the feast day. I decided to opt out. (In the South family, we are allowed to skip the odd Thanksgiving without fear of reprisals, as long as we show up for Christmas.) Instead, I decided to test my fantasies by spending a small chunk of my new salary on a boating experience. I flew to Fort Myers and took a five-day, one-on-one course, learning the basics of how to operate a trawler. At the time, it seemed absurd—pointlessly fun: like taking a cooking class when you can't boil water, or going to car-racing school when you normally take the bus.

I was learning (and staying) alone that week on a single engine 32-foot Grand Banks trawler from the mid-1970s. Even with my limited experience, I could tell it had seen better days. But I loved playing house on it the first night, bringing groceries aboard and cooking, enjoying a drink on the bridge while the sun went down, listening to the murmurs from other nearby boaters enjoying their cocktails. So far, this boating stuff was great, everything I hoped for. Maybe I could get a trawler, live on it and never even leave the dock!

My instructor arrived the next morning and explained the course objectives: I would have five days to learn the basics of the engine and electrical systems, navigation, safety, docking and maneuvering. I'd have to be able to plan an overnight trip, plot it, take us there, anchor in the harbor, bring us back and dock. Yeah, sure. Or then again, *maybe I could get a boat, live on it and never even leave the dock!*

I learned a lot that week, and though I was very nearly overwhelmed by how much there still was to learn, I gathered a vague sense that I could do this. I had gone from reading magazines and picturing myself on a boat to actually running a boat very badly—nervously watching my stern drift out of the channel between markers, messing up running-time calculations, forgetting port and starboard repeatedly. Still, here I was at the helm, soaking it all in, slowly but surely improving and, most of all, feeling utterly thrilled to be in over my head.

Toward the end of my week in Florida, on what seemed to a Yankee an oddly sunny Thanksgiving, I remember sitting alone in perfect contentment, washing down my dry store-bought turkey dinner with a seabreeze and staring at the sun sparkling on the water. And of course, I thought what we all think at some point on a great vacation: *This is the life for me.*

But a few days later, my tan and I were back in the dark and dreary Pennsylvania slouch-toward-winter, editing books, writing flap copy, sitting in fluorescent-lit meeting rooms, drinking too much in the evenings and watching a lot of bad television. (A combination I highly recommend for those wishing to bulk up rapidly. I had already gained 10 pounds in

Pennsylvania Dutch country and *I hadn't even touched the baked goods!*)

Each morning I would find myself sitting at the computer for hours before going to the office, ordering boating books from Amazon.com, looking at Dutch steel trawlers, at French barges, at sailors' web pages and online nautical magazines—at any number of Internet sites that suggested a very different life in a very different place. I've always had wanderlust, a short attention span, a completely unrealistic and fickle sense of what I want. I have contented myself with frequent trips, with elaborate fantasies, with a grand view of what the future could hold...to the exasperation of my very driven oldest brother. My endless interest in the lives I might have led even amuses me. I admit it: I've never been normal.

And yet...I had recently seen a toothpaste ad that made me cry. What I really wanted—like most everybody else—was to climb into bed with the same wonderful person every night and know that my world, wherever I was in it, could happily be reduced to the sound of another's breath, rising and falling. The nice life I had competently constructed for myself was starting to mock me with what it *wasn't*. It wasn't challenging, it wasn't satisfying, it wasn't even important to someone else, let alone me!

My last serious relationship had ended more than two years ago, and even though I had made a real effort to move on, it still haunted me. We remained close friends with an oddly intimate connection to each other. That's probably why I felt a little sad every time we got together—it was a friendship built on the wreckage of love, and it subtly taunted me with how much could be right about something that still didn't work out. All my subsequent stabs at dating or relationships seemed, in retrospect, half-hearted. I didn't mind being alone—in fact, I liked it—but the lingering broken heart had the effect of making me feel lonely and I did not like that.

So there was no doubt that I was just one of those people who craved home, family, a place in the world whose vector is love. But for whatever reason—choices made too quickly, a naive belief that things can always be worked out, a tendency to lose myself a little in my desire to adore someone else—love continued to elude me. Sometimes I wondered if being gay made a difference. You do the math: if 10 percent of the population is gay, and let's generously say that means only 5 percent are lesbians, how many of those are beautiful, funny, smart and available? My results are not scientific, of course, but I'd guess somewhere around six—*worldwide*—and I had already dated four of them.

Perhaps it was natural for me to consider taking my solitude on the road, to hope that the sensation of movement would create an illusion of meaningful destination. It wasn't exactly running away because I knew all too well that wherever you go, there you are. But I thought an adventure might at least distract me from the tiny universe of two that had eluded me. Okay, I admit it. I was a little depressed. And much, much more than that, I was disappointed with myself.

At my friend Holley's apartment one night, I flipped through the album of daguerreotype portraits from the late 1800s that she keeps on her coffee table—something I had often done absentmindedly while we talked and drank wine before her cozy fireplace. Buried between blue velvet covers were sixty or seventy people, mostly young at the time, immortalized in the dour expressions that must have been all the rage back then. To my lazy modern eyes, so caught up in the vividness of my own reality, I had always thought these people resembled each other in their creepiness. But on this night, for the first time, I saw that they didn't. Each one of these faces had been animated, had expressed the joy and grief, contentment and longing, peace and frustration of being alive. They had made their own families, had their own love affairs, heartaches, dreams, plans and disappointments— just minutes ago! It took my breath away.

From then on, the fleetingness of everything we are— blood, bones, brains, dreams, hopes, loves—would haunt me in little gusts—sometimes when I was laughing about a moment that seemed to have happened yesterday but was really twenty-five years ago. Once I recalled a classmate who died at 12—and then I counted the years I had lived beyond him. And I was completely slammed upside the head by it when my grandmother died and The Farm, the only fixed home I had known in my peripatetic childhood, was emptied and on the market within a week—its old farmhouse and weathered barns no doubt destined to be torn down and replaced with a development of vinyl-clad colonials; its beautiful, rolling acreage probably on the way to being subdivided into tiny, tacky parcels and renamed something unintentionally ironic like The Estates at Bear Trap Farm.

Was I having a midlife crisis? The timing was right. But to

me, it seemed more like a reckoning—a complicated concoction of ennui and despair that was nothing more than appropriate. I think most people face this at some point. Some drag it around like an albatross for years. It can be disguised as depression. It can be subdued by drink. It can be pushed back into the corners of our minds by great vacations, by fantasies, by love affairs. But I was no longer able to fend it off.

I suppose I wanted to see if I could resolve my crisis of meaning by living out a wild dream, by casting off the harness that had held my nose to a perfectly pleasant grindstone— keeping my mortgage and car payments current but demanding little from my heart or soul.

So many of us have a secret dream, something we set aside for another day—when the timing is better, when the kids are grown, when there's money in the bank. For me, there was an actual moment, a tipping point, when I stood at the edge of a chasm, just another daydreamer, like you—then threw caution into the teeth of a gale, closed my eyes and jumped.

I was in a New York hotel conference room at a company "offsite" where the main topic of the meeting was the phenomenal success of a book I had slaved over. The heads of sales, publicity, marketing—everyone but the mailroom and food service departments—were all up on the dais, participating in an orchestrated frenzy of congratulation and self-congratulation that was almost as comical as it was icky. It was a little like being forced to watch your colleagues in a very unsexy orgy. *Okay, you do me, then I'll do you and you do him, then he'll do her.* As much as I wanted someone to acknowledge my many months of hard work, I felt even more keenly the need for a very hot shower, Silkwood-style. I was in a contamination zone and felt panicky about getting out.

During the next coffee break, I calmly pushed through the ballroom exit doors as though I was going to the restroom with everyone else, walked past the banquet tables piled with spring waters and caffeinated beverages, through the densely carpeted lobby with its obscenely expensive floral spray and softly piped music, and out onto the hard and dirty midtown sidewalk. I remember standing still, looking up at the blank gray sky and buttoning my jacket against the fall wind. My heart was racing and I felt a prickly, intense flash of recognition. This was the same sensation I'd known in dreams, when I was being followed or chased. I wasn't just skipping a boring late-afternoon meeting. I was *escaping*. I stood on the sidewalk and felt my soul slipping into the crowd, walking fast, faster and finally sprinting away. And that's when I realized I had just quit, that I wasn't going back, I was done and utterly free.

This is the most honest explanation I can give for why I woke up one day, a 40-year-old book editor with virtually no nautical skills, and decided to throw away my old life, buy a boat and go to sea. *Fortes fortuna adiuvat* was our family motto. And Fortune *does* favor the brave but—let's face it—I also had nothing to lose.

CHAPTER TWO

If wishes were fishes the ocean would be all of our desire.
—GERTRUDE STEIN

So, I was going to follow my salty bliss.

Once I'd made up my mind, it all began falling into place. Actually, it was sort of like pushing a boulder off a cliff: after the first shove, everything else seemed inevitable.

The house went on the market and I quickly found a buyer. I started sorting through my belongings and paring things down for life aboard a boat. Oh, yeah: a boat, I still needed to find a boat.

My obsessive online wanderings, previously symptoms of a fantasy life gone awry, were about to pay off. Although I still had plenty to figure out, it was absolutely amazing how much I'd already absorbed. And if I knew only one thing (though some would call that an overestimate), it was that if I was going to sea, I was going in a trawler.

There has always been a great divide between those who

motor and those who sail. I could try to delineate it for you, but it is probably best compared to the ancient schism between those who wear briefs and those who wear boxers, or those who cook and those who bake. It's personal, essential and somewhat mysterious. Aesthetically and spiritually, I had always been drawn to the romantic simplicity of sailboats. Yet I recognized that being a great sailor takes not just a thorough knowledge of basic seamanship but a whole new vocabulary and years of understanding the wind. Mystique aside, it's a much more complicated proposition than running a powerboat. My respect for the sea, as well as my innate laziness, left me with no doubt that a powerboat would be easier to master—or at least skipper with competence. And since I loved the lines of a salty-looking workboat nearly as much as a classic wooden sailboat, a trawler it would be.

Trawlers, which were originally fishing vessels that trailed nets, come in many sizes and shapes—for instance, the rusty shrimp and scallop hunters off the U.S. coast; the large European, Scandinavian and Baltic boats that catch tuna, mackerel and anchovies, and even the ruggedly adorable crabbers off the British Isles. But you probably know them best as the gaily painted wooden souvenir miniatures, sold in every seaside town from Apalachicola to Wellfleet. Of course, these kitschy tributes to a town's fishing heritage, usually painted bright blue or red and adorned with minuscule lobster traps or nets, have as much in common with their real counterparts as George Clooney does with a real captain of a swordfish boat. They're mere Lilliputian replicas of the R-rated originals that go to sea and stay at sea until the holds are full of fresh catch, no matter how rough the weather.

Somewhere in between these two extremes, you will find modern trawler yachts. Made of fiberglass, wood or steel, they may have a single engine or twins and their fuel capacities and ranges vary wildly. Their looks are deceiving, too: some of them are the spitting image of their blue-collar relatives, and some resemble the love child of a fling between a naughty tug and a slick cigarette boat.

True trawlers always have one thing in common: a full displacement hull shape. A displacement hull is just what it sounds like: the bow literally plows through the water, smoothly sweeping it aside as it makes way. This form of travel is slow but economical, and it's the most important reason (of several) that trawlers are often capable of ocean crossings. While their working-class cousins toss relentlessly in 15-foot swells for weeks on end, slowly depleting their fuel tanks as the fish fill the hold, today's trawler yacht has adapted the same slow engine and fuel economy to the purpose of exploring the world in comfort and safety.

In the last few decades, recreational trawlers—once spurned by sailors as little more than houseboats because of their engines, amenities and limited coastal range—have undergone a kind of revolution. That's due in large part to the development of smaller trawlers with ranges of over 3,000 nautical miles, built to withstand any kind of sea conditions. These vessels are as capable of circumnavigation as a sailboat yet far more comfortable and reliable—that is, dry and independent of the wind's whimsy. They are mini-ships with salty pedigrees that lure lifelong sailors into warm, comfortable pilothouses as a form of luxurious graduation rather than sissy shame.

My needs were basic: I wanted a boat that was handsome, fuel-efficient, and most of all, seaworthy. The catch was, I had to be able to *afford* it—and that very conveniently ruled out 99 percent of the trawlers on the market. In other words, I was not looking for a boat but for a miracle.

Every morning I logged on to www.yachtworld.com (and several other sites) and perused their thousands of listings like a woman possessed. The pickings were slim. If I could afford it, it was thirty years old and a call to the broker inevitably revealed expensive "issues" that needed to be addressed before the boat was fit to splash. If it was absolutely perfect, it was generally a half million dollars more than I could dream of spending. At points, I became so discouraged that I started considering bigger compromises. Maybe a motor sailer. Maybe a classic wooden sailboat. Or maybe just a very large inner tube.

One day when my efforts were starting to seem hopeless, I tried entering something different in Yachtworld's search-engine fields. I had been looking at Nordhavns, Krogens, Fishers, Cherubinis; at steel, fiberglass, wood; at sailboats, motor sailers, trawlers. But on this morning, I haphazardly tried out the word "custom," and a secret cyber-wall swung wide open, instantly revealing a dozen listings I hadn't seen before. One of them sent me into immediate orbit: a 40-foot custom steel trawler in Pahokee, Florida.

At first glance, the boat seemed way too good to be true.

Shady Lady was only thirteen years old. Photos suggested that the interior was positively spacious—and good-looking, in a utilitarian way. (Even a lot of the luxury trawlers have interiors that look like fancy RVs or tacky 1980s condos.) The listing details claimed two staterooms, two heads, a pilothouse, a big salon with a galley in the corner, a walk-around engine room with workbench (virtually unheard of on a 40-foot boat) and plenty of outdoor deck space. Fuel capacity was 750 gallons, which gave this boat a cruising range of over 3,000 nautical miles. Its tanks held 400 gallons of water. And it was at the high end of what I had decided I could afford—roughly one-quarter of the price of a used 40-foot Nordhavn.

A call to the broker revealed that *Shady Lady* had been on the market for a few months and that the owner was also the builder. A master steelworker, Mel Traber had built the boat for his retirement, with the design assistance of the legendary Phil Bolger. Since I'd become a fanatic researcher, I already owned a copy of Bolger's book, *Boats with an Open Mind*. I rifled through the index and found *Shady Lady* on page 392. She was featured as an example of a rare trawler design by Bolger, whose cult following consisted mainly of sailors.

I saw only two immediate drawbacks to this boat. I had originally hoped to find a vessel capable of circumnavigation—not that I was deranged enough to attempt that, but I liked the *possibility* of it. Bolger's text revealed that *Shady Lady* was designed for going as far offshore as Bermuda, which is about 600 miles out. Unless I added paravane stabilizers (the large outriggers that you see on many fishing boats), she would roll a bit too much for the kind of continuous and serious swells an ocean crossing might entail. She also lacked other equipment that would make her ideally suited for a transatlantic crossing: a backup (or "wing") engine, a generator, a water maker. Most of this could be added if I had the money but for now, I would have to limit myself to coastal cruising if I bought this boat.

My other hesitation was that despite liking almost everything else about *Shady Lady*'s lines, I had some aesthetic concerns about the stern. In the small online photo, which was difficult to see, its rear end looked big, high, square. Lots of junk in the trunk. Bootilicious. Packing much back.

I tried gently quizzing the broker about this but it was hard to be subtle.

"Ummmm. I really like the looks of this boat—it seems great—but the...ummm, stern. Is it kind of...ungainly? Boxy? Ummm, I guess what I mean is...butt-ugly?" His stiff reply was, "I don't know. It looks fine to me," in a tone that implied I must have some kind of sick derriere fetish to even notice such a thing.

I knew I needed to get down to Florida right away and have a look but my house closing was just a few days away and I was only about halfway through my packing. Though I was a veteran itinerant, who had never lived anywhere for more than two years until I moved to New York City in my midtwenties, I loathed the process. Usually, I got away with just jamming everything into trash bags at the last minute. But since I was moving aboard a boat, I tried to embrace the minimalist fantasy wholeheartedly. I had visions of myself with nothing but a couple of pairs of khaki shorts, a closet full of crisp white button-down shirts, no more than two pairs of worn TopSider sneakers and an array of baseball caps that would take my outfit from daytime casual to...well, nighttime casual.

But it was just a fantasy. When push came to shove, I found I was appallingly attached to nice things. I couldn't part with a couple of designer suits, even though I hoped I'd rarely have to wear them again. And though I'd already read my many hundreds of books, I couldn't discard them. (Me parting with my books was like a mechanic parting with his wrenches, a chef parting with his knives, a soccer player parting with his legs. As a seasoned editor, I also knew these awful analogies were indicative of my desperation to justify keeping the books, and that seemed reason enough.)

Rising above my fondness for the Hefty-bag method of packing, I carefully filled several plastic boxes with things to make my trawler feel more like home: all my nautical books, my favorite kitchen stuff, fancy glasses, an Hermès tray, a silver ice bucket, some framed family photos—all these were separated, wrapped and marked BOAT. I even packed a fancy electric espresso maker *and* a drip coffeemaker. (Shows how much I knew: unless you have a generator aboard, only 12-volt appliances work when you're away from the dock.) It was psychologically tougher to let these material things go than I had ever expected: they were the small rewards I'd provided myself as compensation for my wage slavery, and I clung to them like a life raft while I bobbed between the two shores of present and future, home and boat.

I had infrequent but intense moments of feeling I was in way over my head. Despite my nomadic childhood (or maybe because of it) I had always felt a deep attachment to at least *a sense* of home. I remember walking through my loftlike living room with its high ceilings and exposed stone walls, looking at the stacks of books on the floor and my art leaning against the walls and feeling a sense of panic. Sure, I hadn't liked it here, but I had made it a lovely residence. All I knew about the next one was that it would float—and that wasn't much to go on. I had no job, I was about to have no house, I still hadn't found a boat. I had jumped into all of this without any kind of backup plan. Every other accomplishment in my life had been part of a sane, linear progression. Now I faced a series of unknown what-ifs. What if my house deal fell through? What if I couldn't find a boat I could afford? What if I couldn't handle a boat? What if I got sick or ran out of money? All I could do, I realized, was surge ahead, clear one hurdle at a time, and keep on believing that I would be okay.

It wasn't much of a game plan, but it was what I had. And so, two days before my brother Tom was scheduled to help me with what was bound to be a nightmarish move, and three days before my closing, I got on a plane to see *Shady Lady* in Florida. It was clear that I would probably not get much sleep over the next four days if I planned to get everything done. But maybe, just maybe, this was my boat.

SKIP, THE MARINE BROKER, picked me up at Palm Beach International Airport, and we headed out to Pahokee, which is on Lake Okeechobee, about 45 miles inland. As we headed west, the endless strip of hot, white, palm-fringed highway and fast-food joints gave way to orange groves and flat farmland that was virtually uninhabited.

Pahokee itself, or at least what I saw of it, was just the way I'd pictured central Florida. Lots of ranch houses with jalousie windows, trailer homes with hurricane shutters, and small cottages with front porches that had long since lost their paint. I had the distinct feeling that people ate black-eyed peas and played the banjo out here. Men in overalls sat on front porches in rocking chairs, waving away flies while hounds slept at their feet. Okay, maybe I made that last part up. But I was pretty sure a girl could buy moonshine in this part of the world. Looking around, there was no way to miss that this piece of Florida, away from the touristy coastlines, was still the deep South.

I caught a glimpse of Lake Okeechobee flashing like tinfoil beneath a bright sky. It was staggeringly big, hence its imaginative name, derived from the Seminole Indian words for "big" and "water." The second largest lake in the United States, right behind Lake Michigan (the other Great Lakes are shared with Canada), Lake Okeechobee is better known today as "The Bass Capital of the World." Less than 15 feet deep, it has a circumference of 150 miles and covers an area of 730 miles, or almost half a million acres. On this hot, hazy day it was impossible to see a shoreline, and its flat, naked expanse stretched out, shimmering, like a bright dead sea.

It was probably just as well that I didn't know anything about my family's connection to the area until days later, when I had returned to New York and was telling my brother Hamilton about the trip. "Oh, Lake Okeechobee. That's where Hamilton Disston did all that drainage." He was vague on other details, but a little research revealed that this part of the world had been well known to my ancestors—actually, more like *owned* by my ancestors. In the 1880s, Hamilton Disston (my great-great-grandfather's cousin and best friend and the man for whom my grandfather, father and brother are all named) became obsessed with the idea of draining the Everglades. A wealthy Philadelphia toolmaker, Disston bought 4 million acres from the state of Florida for 25 cents an acre, instantly becoming the largest landowner in the United States, with over 6,000 square miles of Florida to his name. Yet today, when I visit the Sunshine State, I am forced to stay at a Days Inn and eat at Denny's like everyone else. It seems so very, very wrong, doesn't it?

Disston, who became known far and wide as the Drainage King (eat your heart out, Michael Jackson!), dredged canals connecting Lakes Kissimmee, Hatchineha, and Tohopekaliga. He also deepened and straightened other lakes that formed the headwaters for the Kissimmee River. He blasted out the waterfall of the Caloosahatchee River and connected Lakes Bonnet, Hicpochee, and Lettuce by canal systems. Disston's projects drained a total of 50,000 acres, increased agricultural lands and created a navigable route from the central Florida town of Kissimmee to the Gulf of Mexico. He also established a large sugarcane plantation in Osceola County and founded the resort town of Disston, which is now known as Gulfport. Despite these successes, the panic of 1893, the repeal of

Grover Cleveland's sugar-growing incentives and several freezes in a row brought Disston's development dreams to a crashing halt. Though he was officially reported to have died of heart failure, it was widely known that he shot himself in the bathtub of his Philadelphia mansion.

I was amazed that all of this had been completely unknown to me. Finding my boat here, in the unlikely place of Pahokee, now seemed like more than geographic coincidence. It seemed fateful. Rather than focus on the billions of dollars that 4 million acres of Florida would now be worth, or the lasting ecological damage my ancestors had ignorantly wrought, the eternal optimist in me decided to see this coincidence as a positive omen about the boat and a rare chance to feel like a relative financial success. After all, the $28,000 I had once drained from my own 401(k) now seemed like a small drop in Big Waters.

We reached the marina, and Skip pulled his new SUV into a parking lot beside a flotilla of other SUVs and American-made pickup trucks. When I opened the door the crisp chilliness of the cab instantly wilted in the midday heat. We walked down a short slope to the no-frills dock, locked behind a chain-link fence and gate.

And there was *Shady Lady*. She looked all wrong for that spot—too distinctive, too majestic, way too salty. I experienced an immediate joy that overpowered all common sense. I hadn't even been aboard yet—she might be a disaster. But something in me knew right away that this was my boat.

The first thing I did was walk to a nearby finger pier and gaze across at her stern. It was all right. Not svelte but certainly not the clunky eyesore I had feared. While the broker unlocked the boat, I climbed aboard and checked out the decks. They were white and almost blinding in the harsh noon sunlight. There was plenty of space at the bow, which was high and solid-looking. Side decks with hand railings stretched back to the stern, which was big and open, with room for a table and chairs. You could have a dinner party for six back there and still have room for a wandering mariachi band. There was even more deck space above the salon, where a hard-bottomed dinghy lay, lashed to one side.

I couldn't believe my luck. I had all but given up on the idea of outdoor space as I researched trawlers. Even the very expensive ones had tiny rear cockpits—on smaller boats designers almost always preferred to utilize every square inch to maximize the accommodations.

Much as I might want to be sunbathing, the pilothouse would be where I'd spend most of my time underway and I loved *Shady Lady*'s. It was shippy, with a forward rake to its big view. The instruments were all aligned overhead on a cleverly hinged shelf that folded down for access to the wires at the back. The electronics weren't new and they weren't fancy, but there seemed to be plenty of them. It was hard to say what was missing since I couldn't even identify most of the equipment. But I liked the white chart table that was topped by a cabinet with four mahogany-stained flat drawers for holding paper charts. There was an upholstered bench with a toggle switch on the armrest that Skip explained you could use like a joystick to maneuver the boat around crab pots and other obstacles without having to get up and adjust the autopilot.

The helm was a stainless steel wheel to starboard, just above the steps to the salon.

If the pilothouse had excited me, the salon left me speechless. There were seven 23-inch portholes with tempered glass and aluminum bolts and hardware. (Most boats I had seen had portholes of 9 inches or less, if they had any at all. The trend was toward larger, squarer, picture windows.) The interior felt bright and spacious, with more than 6½ feet between the cabin sole and the painted steel beams that ran overhead. Usually, interiors turned out to be smaller than they looked in website photos, but I was amazed by the size of the salon. In the forward port corner was a galley with a stainless steel sink, a small wood-topped cutting counter and a full-size gas stove. A big cabinet with reach-down refrigerator compartments bordered one side, and perpendicular to that was a long countertop, with storage underneath.

Behind the counter was the sitting area: nothing fancy—a varnished trestle table with a matching bench on one side and an upholstered settee on the other. The white surfaces with dark wood drawer fronts and trim (known as "Herreshoff style" in boating circles) continued throughout the boat and did a lot to keep things cheerful.

At the forward end of the salon, a short flight of stairs led down to the guestroom. It had a double berth and a small head on the port side and a single berth on the starboard side. Overhead was a big square hatch that propped wide open for sunlight and air. The portholes down here were somewhat smaller but still very big.

If you turned and faced the stern, you were in front of another doorway with metal steps that went down into the engine room. Because *Shady Lady* had a box keel, the engine sat down very low, providing extra space and stability. You could walk all the way around the engine, which was a very basic Ford NorEast 135-horsepower diesel. Of course, I knew absolutely nothing about anything in this engine room at the time and planned to have an expert survey it for me. In the meantime, Skip showed me all of the well-thought-out details and explained that this was an excellent diesel to have because parts were readily available for it anywhere in the world. Everything on this boat seemed to be about simplicity and good design.

I had looked at enough boats to know that the average engine "room" was hidden under a hatch in the salon sole. Once you pulled up a bunch of heavy floor panels, you faced the unappetizing prospect of crawling down into a tiny dark hole with a big hot engine and no room for maneuvering. That was what an engine room looked like. This one was bigger and brighter (and probably cleaner) than my first apartment in New York City.

We passed back through the salon and down the steps to the master stateroom in the stern. On the starboard wall was a white countertop above a series of built-in varnished drawers. There was a queen-size bed with room to walk around on either side. On the wall at the foot of the bed was more storage: two dark varnished cupboards on either side of a matching bookcase. A countertop and sink lined the port wall, with a toilet tucked away behind a small privacy wall. At the other end was a steel shower stall with a built-in bench. Big portholes everywhere.

I was beside myself. This boat appeared to have almost everything I wanted, even though I had long since concluded that it was going to be impossible to find *at any price*.

I shot a roll of photos and then Skip locked up. We sat down at the small lakeside concession stand, and I started filling out the necessary paperwork for making an offer. I didn't even need to think it over. What I knew about boats was very little, but I had fallen in love at first sight. This was my boat. I had to get it. There would be a survey before our deal closed, and that would give me an opportunity to back out if my beloved was revealed to be a crazy waste of money.

Skip had some fried conch and a Heineken as he walked me through the forms. I was too wound up to eat, but I was not about to stand by and witness the tragic spectacle of a man drinking alone. My beer was icy cold, the day was swelteringly hot and I was happy as a clam.

I flew home, and two days later, my belongings were gone, I was no longer a homeowner and I had made a deal on the boat of my dreams.

PEOPLE OFTEN ASK ME, why *this* adventure, why a boat and a life on the water? There's a popular belief that those who go down to the sea in ships must do so because they were born to it—or because they were exposed to it so young that they caught it, like some kind of virus.

In my case, I have no single, logical explanation, though I can offer up a host of coincidences. I have lived near the ocean, off and on, throughout my life. I crossed the Atlantic by ship several times in my youth. My first crush was on a Russian sailor and my second on an Irish fisherman. And then there's blood: my grandfather's brother was an admiral in the navy. And my grandfather, who retired as brigadier general in the Marine Corps, was the naval attaché to Brazil for a time and, after his retirement, a public relations director for the famed Moore-McCormack ocean liners. Hard to say what's cause and what's effect. But I often find myself marveling at the intersection of (an often unknown) past and the present in my life. Finding my boat on Lake Okeechobee is an example. Some people call this coincidence, or synchronicity, or serendipity. Call it what you will, but the older I get, the harder I find it to believe that anything is entirely accidental.

And it was comforting to remind myself of this belief as I waded into the grueling process of making the *Shady Lady* mine. Though the owner and I came to an agreement on the price very quickly, I was unprepared for how difficult it would be to get financing. I had about 50 percent of the boat's purchase price to put down in cash, but I immediately ran into other obstacles that threatened my deal.

First, I went to a specialized marine broker, who shops for money for boat purchases much the way a mortgage broker does for home buyers. She was kind but very discouraging. The fact that the *Shady Lady* was not a production boat but a custom build was a big strike against her in a lender's book. Boat lenders, like banks, look for "comparables" when you're applying for a mortgage. Obviously, if you're not getting a brand-name boat that thousands of other people also own, the lender doesn't have any similar boats to evaluate it against.

This prejudice was a frustrating discovery because the *Shady Lady* was as seaworthy as they come—her steel hull was much less prone to damage in collisions with rocks, docks or other boats than a wooden or fiberglass vessel. And as long as you're willing to fight a tireless crusade against rust, steel is remarkably impervious to the harsh home offered by the sea. That's why almost all commercial ships are custom-made of steel. But, let's face it, common sense is anathema to the bureaucratic decision-making process.

Last but not least amongst my prospective stumbling blocks, the broker warned me, boats are considered luxury purchases, and because they are mobile, lenders require very high, if not perfect, credit scores and substantial extra assets in the bank as additional assurance that you're not some fly-by-night. With my collateral assets sold and my new self-employed status, I was pretty much a boat lender's worst nightmare. She suggested I go directly to my local bank. I felt like a patient who has just been given a specialist's phone number in sympathetic tones that implied "Poor thing. Let her at least go through the motions." I suspected the diagnosis would be terminal.

It was.

Now I was worried. I was running out of options—I might have to let this deal go and start over. I knew, too, that the kind of boat I could buy for half the price of the *Shady Lady* was going to be much less than half the boat.

With absolutely no confidence at all, I told the boat broker of my difficulties and asked if the owner would consider holding the loan for me, in return for 10 percent interest. (With my usual economic savvy, I pulled the 10 percent number out of thin air—it seemed worth his while but was still less than what a major bank might charge.) Much to my surprise—and immense relief—he agreed to my proposal and we set a closing date.

But that didn't mean my troubles were over. Oh, no. There was still the survey to come, and then I had to get marine insurance. The fun was just beginning.

CHAPTER THREE

Smooth seas do not make skillful sailors.
—PROVERB

A month before my boat closing, and a week after my house closing, I returned from a short, sunny vacation in Brazil to a late March snowstorm in New York. The dirt road to my brother's Connecticut barn where all my belongings were stored was impassable, which meant that all of the careful packing I had done for the boat had been pointless. *That* would teach me to deviate from the trusted Hefty-bag method.

Armed with only the duffel I had packed for my beach getaway—which contained a pair of sandals, a pair of moccasins, a bathing suit, suntan lotion, two pairs of khakis, two sarongs, four men's button-downs, an assortment of baseball caps and several cups worth of sand—I piled my two trusty canine companions, Samba and Heck, into the car and we left for Florida. My classes at the Chapman School of Seamanship would start in a week, and I needed to get down there a few days before that to find a temporary apartment and rush the boat closing along.

The drive south was more fun than I'd expected. It was exciting to be watching my old life wind away in the rearview mirror as I sped toward the beginning of my new one. I made the trip in two days, with overnight stops in North Carolina and Georgia. The I-95 corridor was bleak, though. Every exit trumpeted its dreary offerings of Comfort Inn, Motel Six, Red Roof Inn, Denny's and McDonald's. The food was always lousy, the coffee was watery swill, the ambiance was pure plastic with a Muzak overlay, the rooms were always dark and airless with low ceilings and ugly mauve bedspreads.

The irony, of course, was that I never went to these establishments, so being forced to frequent them was a form of exotic, if cheerless, travel for me.

Not that I am a snob—at least, not in the traditional sense. I grew up in a family with great taste and very little money. In tenth grade, all I wanted was to live in a vinyl-clad split-level with wall-to-wall carpet, like everyone else. But the house we moved into that year wasn't much to look at—unless you squinted. Very hard. Three stories tall in the front, with four square towers, the faded clapboard front of the rundown manse was connected to an older, simpler wing from the 1700s. The entire house had faded to the color of a winter sky, but it was strangled in green vines. My parents fell instantly in love with the place, and several phone calls later, we moved in to what was known locally as the Campbell Mansion. The owner's great grandparents had modeled it after their ancestral castle in Scotland. Despite the fact that it had a small fountain in a side yard, a many-mullioned sunroom with a tilting terra-cotta floor, a marquetry-floored ballroom and twenty-four other rooms, I never failed to giggle that year when someone referred to our house as "the mansion." It had character, for sure, but the floors all sloped. In the winter, we sealed off most of the house and stayed in the back wing, where the inside of the windows were often covered with frost.

My dreams of suburbia died a slow, bitter death after the first three days at the new place. The entire family spent this time hacking away the ivy from the windows, so that the kitchen could see light for the first time in decades. But, I have to hand it to my parents—within a month we had refinished all the floors, painted all the walls white, hung art, unboxed books and rolled out the threadbare orientals. And it was beautiful.

My friends, most of whom lived in stuffy, overheated houses with fake wood paneling, would come to my house where something homemade was always cooking and classical music played, and say with reverence, "Wow. I wish I lived in a place like this. Your parents are so cool."

It was true, in retrospect. Mom and Dad never worried about conforming, and despite an income that made raising four kids quite a feat, they made sure they spent their money wisely: we almost never went to the movies or out to dinner, which other families seemed to do all the time. Instead, we'd be commanded to entertain ourselves. As we got older, we entertained them instead, with art shows and short plays, where we charged a small admission fee (naturally).

In our house, new clothes usually happened once a year, right before school started. I later realized how many of our exotic and delicious meals cost only a few dollars to assemble. Arroz con pollo. Feijoada. Beef stroganoff. Chili con carne. And always a big salad. My parents somehow squirreled away enough to take us all to Europe and back by ship, twice. In fact, the first meal I ever remember eating "out" was the second seating for dinner in the dining room of the SS *Mikhail Lermontov* when I was 11.

So, force me to choose between cheap and beautiful and I will always be impractical. But thanks to my parents, I know that, with imagination, the one doesn't always exclude the other.

This was not a concept that the Studio Six corporation embraced. My efficiency unit in West Palm Beach was about as sterile and dull as they come. But for $75 a night, I had a room with a tiny kitchen where my Jack Russells were welcomed, and when I took them for walks, we stepped on thick bermuda grass lawns, gazed at tall royal palms and basked in the Florida sunshine. It was a fine stop-gap until my housing dilemma resolved itself.

The first day of school fell on the last Monday of March. I drove 40 miles north to Stuart and sat in a fluorescent-lit room with low acoustic ceiling tiles and a clock that ticked like a dripping faucet. There were nautical charts on the walls, and near the window, there was an overhead projector on top of a cart stacked with nautical teaching aids. A desktop lectern at the front of the class displayed a small yellow sign that said: "Leave this classroom as you would leave a boat." (After we had experienced the fleet of teaching boats, we would joke that this *could* be interpreted as an invitation to trash the place.)

There were fourteen of us enrolled for the spring 2004 professional mariner training program at the Chapman School. At orientation, as I looked around the room, I felt my heart sinking ever so slightly. We had just heard a welcoming speech from a school administrator that I had found...well, a little bit bizarre.

A spryly elfin woman with a bowl haircut, she had been instrumental in the school since its founding. After a rambling introduction on the history of Chapman, she segued into a self-aggrandizing speech about her tireless efforts on behalf of the school. With a lot of hard work and exhausting fund-raising efforts—which mostly happened at Chapman as a by-product of selling off donated boats—she had arranged for Chapman to purchase the adjacent boatyard to add to its campus. But not, she told us, before she had made a deal with the Lord: if He would allow her to somehow swing the purchase, she promised Him she would start a boating program for high school students. After this "I'll scratch your back if you scratch mine" arrangement with the Almighty, the purchase happened so easily that it was "eerie" and she felt honor-bound to keep her commitment.

While this speech initially set off some alarm bells about what kind of crackpot school I had committed myself to, at least it answered some of the questions that had nagged at philosophical minds since the beginning of time. Why was the world plagued by disease, hunger, war and injustice? Obviously, God's little-guessed preoccupation with safe boating for high school students had distracted him from the nastier ethnic cleansing situations around the globe. This woman seemed a little bit wacky. So on my very first morning of school, the thought that would haunt me for months to come first flitted into my mind: Had I just flushed $6,000 and nine weeks of my life down the drain?

It was hard to gauge my fellow students' reactions. We were a motley crew, ranging in age from about 18 to 65, most with previous boating experience. The class had only three women, all in our forties. Half of my fellow students were young men in baggy shorts and backward baseball caps, launching careers in the marine industry. The middle-age students were either early retirees about to change their life-styles or...whatever the hell I was. After hearing each class-mate introduce himself, I found it hard to believe we had enough in common to form fast friendships. But that was okay. I was here to learn, and the Chapman School of Sea-manship was reputed to be one of the best schools of its kind.

The curriculum certainly appeared exhaustive. We were about to embark on courses covering marine engines, marine electronics, marlinspike (knotsmanship, essentially), boat maintenance, boat handling, chart navigation, seamanship, emergency first aid, marine weather, boat systems and Coast Guard rules. Classes would start at 8:30 A.M. and would each be two hours long, with a one-hour break for lunch. Though we were scheduled for three classes per day, we almost always had study hall or an extra class after that, so our school day really ended at 5:30—unless you counted homework.

It had been a long time since I had sat in a classroom, but it never occurred to me to worry. I had always been one of those students that the hard workers hated, with reason. Good grades came easily to me. My very lax study habits and very lucky retentive abilities no doubt developed as a by-product of my quirky childhood.

When I was 13, my parents sold all their belongings and moved us to the west coast of Ireland. Dad was an artist who had been teaching studio art and college-level art history for years. Mom was a poet and sometime medical researcher. They both wanted a chance to do what they loved for a few years, so off we went on the grand adventure.

We lived in a big house called Walker's Lodge, an austere place way out on its own small peninsula, with no running water or electricity. Every day my brother Hamilton and I would walk a mile around the bay to the bottom of a paved road, where we'd change into our school shoes and hide our Wellingtons in the blackberry bushes. At the top of the road, we caught a city bus into Sligo, where Hamilton attended the Christian Brothers' Summerhill School and I went to the Ursuline convent school.

About halfway through my first term there, I was moved up a year, and into an honors class. I was getting an amazing education at the Ursuline: I took classes in Irish, Shakespeare, French, religion, poetry, but the move from one year to the next was disastrous for my future in math. I jumped suddenly from beginning algebra into the middle of trigonometry. It was like being hit on the head with a frying pan, and my mathematical faculties never recovered. In fact, I developed a kind of math amnesia; my head still hurts when called upon to perform anything more than simple subtraction or multiplication.

After a year and a half in Walker's Lodge, my mother decided to go to graduate school at the London School of Economics, so off we went to London. My brother and I were now enrolled together, at the Holland Park Comprehensive School, which was as different from our Irish Catholic schools as you can imagine. HPC was (and still is) somewhat notorious for its liberal curriculum and multinational, educationally egalitarian approach. Immigrants from every nation on earth, with wildly different language skills and educational foundations, shared classrooms where each student was supposed to learn at his own pace. It was unstructured and chaotic, but my creative impulses flourished at Holland Park. I played the flute in a jazz band. I started writing funny short stories. I also narrowly avoided getting knifed in the schoolyard by a tough Nigerian girl who heard I thought I could beat her up. (It turns out my younger brother had told her younger sister that this was so. Fabulous.)

At 15, when I returned from these two years abroad, I found myself at an upstate New York high school where my fellow tenth graders were (I kid you not) reading a book about a raccoon named Ricky. Luckily, my English teacher noticed me slipping into a coma at the back of the room and sent me to the library for the rest of the year to write a paper on James Joyce. (On the other hand, I was the only person in my class who almost failed geometry.)

The following year I was a Rotary International exchange student in Brazil, and when I returned, I skipped my senior year of high school and commuted to a local college. I wound up at three colleges before I enrolled at the last one, Manhattanville College, where I had an academic scholarship. In a desperate attempt to make up for some credits that didn't transfer from an Irish college, I carried thirty-two credits my last semester. It wasn't an easy spring, but I couldn't afford another semester. Somehow I managed to get through it without dragging my grade-point average down too much.

While I may have gone into Chapman with a certain academic nonchalance that bordered on cockiness, it quickly became apparent that I was not going to coast through these nine weeks. I was no longer in the domain of the humanities, where years of being a fanatic reader prepared me for anything. All the material I studied was completely unfamiliar and was presented at a breakneck pace. The days were incredibly taxing. I'd struggle valiantly with set and drift calculations for two hours in Chart Navigation only to find myself completely baffled by the process of heat exchange in the next hour of Marine Engines. I usually finished a day feeling overwhelmed and dispirited— I knew that if I didn't stay on top of reviewing each day's lessons as well as the homework, I wouldn't make it.

I was not alone. Most of our class quickly realized that this course was something we just had to soldier through. It was a shame, really, because there was nothing intrinsically boring about most of our subjects.

There was an enormous emphasis placed on preparing us to pass our Coast Guard licensing exams. These are the same exams that students at merchant marine academies take, so in essence, we were cramming four years' work into nine weeks. Also, the nature of what the Coast Guard wanted us to know and what competent boaters need to know are not one and the same. The exams themselves are fiendish, riddled with information designed to point you toward the wrong answer, and the entire process seems more like running a gauntlet designed to test your endurance, memory and masochistic tendencies than it is meant to measure your maritime knowledge.

And that was galling because, deep down, as lazy as I am, I like learning new things and I had hoped I'd exit Chapman a wizened old salt, replete with a parrot grafted to my shoulder and a tendency to sprinkle "Avast, ye hearties" throughout my conversations.

While passing the Coast Guard tests was important for anyone who needed a captain's license, the core of the older students' frustration existed because we had come to Chapman to *learn*, not just to pass a test. It was also pretty obvious that we Professional Mainer Training students were not the school's top priority. The electronics lab, which had a big Raytheon sign overhead, was full of equipment that was twenty years old. Those components that did function were completely obsolete. It was like going to business school and being asked to calculate profit-and-loss reports on an abacus.

The Chapman fleet of practice boats consisted of about a dozen fiberglass sailboats and trawlers, most in the 30-foot range and comically decrepit. They had all been donated, of course; as one instructor noted, there was no sense letting the students beat up good boats. It was a fair point, though the teachers were alert and agile enough to help avert even the softest collisions between boat and dock before they happened.

Any complaints we students made about the inadequate facilities and equipment were met with a routine volley about the nonprofit status of Chapman, as well as expressions of incredulity. I think we were the first class that had been dissatisfied and very vocal about it. But from our standpoint, $6,000 should have offered us experience on at least one boat in good condition with up-to-date electronics. Chapman may have been nonprofit, but our tuition was not—it equaled a year's education at a maritime academy or a semester at many excellent universities. All of us worried about graduating from a supposedly first-rate seamanship school without really knowing how to work a global positioning system, for instance. It was great that we'd understand dead reckoning and how to take a three-point bearing, but out in the real world, a GPS was standard equipment that we'd use day in and day out.

On top of the frustrations of school, I was still mired in the process of buying the boat, an experience that is a lot like buying a house in terms of hassle and paperwork. The survey of *Shady Lady* revealed some shallow pitting in the steel below the water line, and that necessitated a reading by sound gauge, which meant paying to have the surveyor come back, blasting all the paint and primer off the bottom, taking the reading and then repainting—a hideously expensive process that devoured the last of my small boat-improvement fund.

Just to dramatically heighten my already soaring anxiety level, I very confidently rear-ended my convertible in the third week of classes. I had missed a turn and was doing an impatient and exasperated three-point turn in a completely empty parking lot. I looked at the emptiness over my right shoulder, threw the car into reverse and hit the gas, slamming into a concrete lamppost that was invisible on my far left. The trunk buckled, the bumper collapsed, the rear quarter-panel had to be replaced. The garage estimated it would take three weeks to repair the car and cost my insurer about $5,000. I would be punished for my carelessness by a huge rate increase and the indignity of having to drive a royal blue Chevy Cavalier in the meantime. The cosmetic work actually took more than five weeks. Hard to believe, but there were no Saab quarter-panels in the whole country. One had to be shipped all the way from Sweden. After week four, when I pointed out that I could have swum to Sweden and back by now, they gave up and just repaired the piece instead of replacing it. Oy vey.

And then I had to make a quick trip home.

A LOW, LOW POINT—NOT just in my year, but in my entire life—was my grandmother's funeral in May. I took two days off from school, flew into New York City, and rode to upstate New York with my brother, who had a car and driver. He rolled business calls on his cell phone the whole way up—not because he wasn't sad and anxious, too, but because everyone has different ways of disguising it.

We all gathered on a hillside, in a tiny cemetery not far from The Farm. Amazingly, everyone was there, not that ours is a particularly large family. But my aunts and uncles and cousins and parents and brothers had come from far and wide on that beautiful day to say good-bye, at last, to Ros.

She had died in January, less than three months after finally agreeing to go to a nursing home at the age of 85. I went to visit her there at Thanksgiving. She was being a good sport about it, but Ros was way too much of a snob to ever make friends in a place like that. Her small room was pretty awful, clean but sterile, and devoid of any personal effects, save a couple of cards, some flowers, and several tins of wintergreen Life Savers, which she consumed voraciously—a pale stand-in for easy access to liquor and cigarettes, I suspected.

Ros had had one hairstyle my whole life: long white tresses pulled back in a loose bun. It was all gone now, chopped into an artless bob that alarmed me and seemed like the physical iteration of her statement that "a priest told me there was nothing sinful about praying to die." This was the hairstyle, the room, the good-bye of someone who wasn't planning to stay much longer, and I remembered with a pang what Ros would often say, back when she was young enough for it to seem funny: "Old age is a shipwreck."

In my youth, The Farm was the center of the universe. It is still one of the prettiest places I've ever seen—a colonial farmhouse with a wraparound porch, perched on 125 acres of stunning land and surrounded by a spattering of barns and outbuildings painted in the flat and faded blood reds of a Wyeth farm scene. The worn brick floor in the kitchen undulated like the surface of the ancient chopping block; copper pots, old baskets, ship models, books, art, the constant sound of many old pendulum clocks swinging time away and into eternity—*good-bye, good-bye, good-bye*—and across it all a swathe of summer sunlight, the liquid amber that surrounds my childhood memories.

My parents were very young then, as were my aunts and

uncles—even my grandmother was still in her fifties. The grown-ups would drink vodka-and-tonics in the early summer evenings and we kids would play outside well past dark. Once in a while, they seemed to forget all about dinner, bedtime, us. They'd roll up the fragile zebra- and lion-skin rugs and dance until the wee hours of the morning. My parents were great dancers, and it felt like something illicit for us to peer through the windows at them, to spy on their sexy, secret life of getting down to Motown. They were just a little more than half the age that I am now.

We were a real family in those days, together for every holiday. But everything changed, of course. Over the years, after my grandfather's complicated and painfully drawn-out passing, my grandmother's drinking increased, just as her own spirits sharply decreased. She was manipulative, lonely, irrational. Her children quarreled among themselves about responsibilities and money, and everyone drifted apart.

Now we stood on a thawed hillside in May ("the month of Mary," as Ros would say, coming through the kitchen door with an armful of daisies and lilacs). My middle brother, Padraic, was driving up from Long Island with his family, but he hadn't arrived yet. We all stood a little awkwardly on the breezy green slope and waited, catching up with each other— veritable strangers after more than a dozen years of distance. I only have five cousins, but I barely recognized them now. One was a lawyer. One was a landscape architect. Another was married with two kids, living in the Adirondacks and teaching school. The baby of the family, who was eternally 6 in my mind, was now a beautiful young woman in her late twenties, getting a Ph.D. in marine biology. It was hard to believe how old we all were and how little we knew each other.

Yet I felt a strong sense of family, of blood being thicker than money, quarrels, time. Whether I saw these people or not anymore, they were mine, part of me, and loving them seemed as involuntary as breathing.

It was hard to accept that it was larger-than-life Ros, with her throaty voice, Bacall-like glamour and naughty stories, packed in a small cardboard box before a deep square of space in the ground. A clergyman stepped forward and said the Lord's Prayer, and I almost beat Ros into the grave when he concluded with "For thine is the kingdom, the power and the glory, for ever and ever." Ros was a devoted though reprobate Catholic and I simply could not believe we were listening to the Protestant version of the prayer. Ros was the kind of old-fashioned Catholic who was appalled by folk guitar services, who got misty-eyed remembering the good old days of Latin mass and the Inquisition. She liked her priests to be like her God: stern, distant, absolute. None of this touchy-feely, street-clothes stuff. It seemed like a slap across the face that she was being commanded to heaven in a language *her* God couldn't hear.

And that was that. No one had prepared anything special for the day. There were no spontaneous recollections, no carefully selected poems, no music, no reception, not even a pretty urn to disguise the flimsy container. After the ten-minute ceremony, we all got in our cars and went home. I was horribly depressed. I thought Ros deserved better, no matter how much she had hurt her children in her waning years. Perhaps they, too, were unprepared, caught off guard by the finality of the moment and the fact that this small cardboard box was what their domineering mother had been reduced to.

I felt like my beautiful childhood had been cremated and scattered in the wind.

THERE WAS MORE THAN one moment in that nine-week period at Chapman when I was stretched so thin emotionally, physically, spiritually and financially that I really thought I might have a nervous breakdown. I didn't feel my usual optimistic, resilient self. It was a massive understatement to say I was not yet enjoying the new life I'd embarked upon. I felt that I was barely keeping my head above water at Chapman. I had no free time to relax or read or do anything else. I was absolutely broke and, added to that, I now felt that my last tie to a physical place on this planet had been severed. The Farm was gone. I missed my friends and family. I was a destitute, rootless wanderer and it was all my own fault.

One sunny spot on the horizon was my boat purchase. All that stood between me and the closing on the *Shady Lady* was getting marine insurance. I went to BoatU.S. first, where they quickly turned me down for two reasons. One, they would not insure a first-time boater on anything larger than 30 feet. And two, they didn't insure anyone on a boat that was more than 10 feet bigger than the last boat they owned. The second denial seemed like overkill, but I guess if you considered my previous boat to be zero feet, since it was nonexistent, then their reasoning made perfect sense.

Next, I tried a marine insurance specialist. I once again ran up against a steel boat prejudice. Many insurers just flat-out won't insure a custom-built steel boat for the same reasons that lenders don't finance them. This broker did find me one quote, but it was astronomical. I'd done a lot of asking around and had been expecting to pay something in the neighborhood of $1,000 to $1,500 for the year. This quote was for $3,200, and that included a discount for the professional mariner training program.

It was now two days before my boat closing. Desperate, I turned to one of my Chapman classmates who I knew had been a hard-charging insurance executive before she and her husband decided to retire early.

I approached Carol at the end of class. She was wearing her bright red Mount Gay Rum Bermuda Races baseball cap and her hair swung behind it in a high ponytail. She had a green Converse sneaker on her right foot and a red one on her left foot, to help her remember starboard and port. Now, she unfolded a pair of incongruously traditional half-glasses and peered over the quote with a frown. "I hate to be discouraging when you're having so much trouble getting a quote but I just wouldn't do business with this company," she said in her honeyed Alabama drawl. "I'm not sayin' they're bad, but *I've* never heard of them. You want to be sure that you get a company that is actually going to pay out if you have an accident, and I just don't feel good about this one."

Being the take-charge kind of gal she was, Carol got on the phone with an old colleague and explained the situation. It was Thursday. Could he find me a better quote, with a better company, before the close of business on Friday? He didn't make any promises, but he said he'd try. In the meantime, after an exhausting day of classes, Carol sat with me and helped fill out the paperwork so that we could turn it around quickly and give him what he needed to, *maybe*, make it happen. I was worried, I was tired and I was so grateful for Carol's assistance I could have wept.

I got a quote—it was just as much but from a great insurer—and the next morning, when I drove 35 miles south with the top down and the radio on, singing at the top of my lungs, I felt the psychic sunshine breaking through my clouds of anxiety. I was on my way to buy my boat. In a small, nondescript office in an unremarkable strip mall, I signed the papers and relinquished my tiny fortune to become the proud owner of a boat that was extraordinary. I couldn't quite believe I had pulled it off.

From this moment forward, the Chapman experience started to improve. I had a home—a small, floating, tuggy-looking home, but a home nonetheless. Although I was tied to a dock and not yet willing to take the newly renamed *Bossanova* out, I was utterly in love with my ship. I was at the Hinckley boat yard in Stuart, which was not a marina and lacked the luxuries of, for instance, Pirate's Cove just across the Manatee Pocket. There were no phone or television hookups, no laundry facilities, no broadband Internet access, no restaurant or bar or swimming pool. On the other hand, there was also no loud reggae music or obnoxious drunks staggering around on weekends, no constant in and out of charter sport fishermen at all hours of the day and night. In fact, there were only three or four other boats with people aboard in the Hinckley yard at any given time. Most of the vessels were in for service or long-term dockage. At night, it was completely deserted and eerily quiet.

Still, every morning at 7:00 A.M. the place came to life as mechanics, painters, woodworkers, electricians and expediters arrived for work. The deserted yard blossomed into a small, nautically obsessed town every time the sun rose. I knew just about everybody who worked there. True to the spirit of boaters everywhere, they were mostly laid-back and friendly—even when I knew they were running behind on jobs and under enormous pressure. Several of the workers carried biscuits in their pockets, just in case they ran into Heck or Samba, who were greeted with affection and treats every time they went for a walk. I liked it there, living aboard my boat.

As for Chapman, in all fairness, most of the staff knew more about seamanship than I'll learn in my lifetime. And many of the instructors were very generous with their time and eager to see us succeed. I knew nothing when I went in but when I came out, I was crammed full of information, some of it very useful, some of it not, and much of it forgotten as quickly as it was learned. Still, there was simply no way I could be forced to learn so much and not leave there better prepared than when I started. Despite its weaknesses, Chapman gave me a great foundation for real-life experience, and it definitely built my confidence.

My classmates topped off the course with an extra three weeks of prep for the Coast Guard exam, but since I had virtually no at-sea experience, I decided to take the test later, when I had accrued enough sea time to qualify as a captain.

In the meantime, Chapman loaned me the phenomenal Captain Bob Swindell for three boat-handling lessons aboard the *Bossanova*, to compensate for a makeup class in electronics that I missed. Captain Bob, who was a professional tugboat captain, met me at the dock one sunny morning. "Good morning, Mary," he hailed me in a booming voice. "Permission to come aboard, Captain?"

"Permission granted, Captain Swindell." Bob climbed aboard, clipboard in one hand, briefcase in another, polarized sunglasses hanging from a cord around his neck. After a short, mandatory love fest with the two vicious guard dogs, I gave Captain Bob a tour of the boat. He whistled when we were through. "This is a great ship you've got here—my kind of boat."

Captain Bob started me off with the basics: "Okay, Mary. We're going to leave the dock in a little while. What's the first thing we should do?"

"Run through a predeparture checklist." I confidently responded. "Check the oil, the fuel, the batteries, the VHF, the lights and steering."

"Nope," he said with a grin. "Those are the *next* things you do. The very first thing you always do is check the weather to see if you should even think about leaving the dock."

We tuned the VHF to WX and Captain Bob asked for my ship's log. "Well, I don't have one yet, but I'll get one."

"Good. And Mary, you don't need to get a fancy one—a simple spiral notebook will do the job. Have you got one of those? We'll use that for now."

We jotted down the date, our location, the wind speed and direction, and then we recorded the morning's forecast. "Now, we know the weather's going to be fine—let's run through that checklist. Let's see how you do that."

Captain Bob watched me as I did my engine inspections. When I showed him how I moved diesel from my port and starboard tanks to a day tank, he nodded approvingly, then suggested I get another log to keep in the engine room. "That way when you check your fuel-tank levels, you can jot down what you had, what you moved and what you wound up with in your day tank."

Back in the pilothouse, we fired up the engine and then stepped out on the deck. "Now, Mary. You're going to have to back out of this slip. What's your approach going to be? We checked the weather, so we know the wind's blowing northeast at 6 knots. Pretty mild. So, which lines would you untie first and what's your strategy going to be once you're out?"

Captain Bob had a wonderful manner as a teacher. Not too hard (he never made me feel stupid) and not too soft (he never let me feel like I knew it all, either). His whole attitude was one of great confidence in my ability to think. This was more important than I can say. The idea of finally running my boat, of starting the engine and getting off the docks, was fairly terrifying to me, despite my nine weeks of intensive schooling. That had all been theoretical. There was nothing remotely abstract about maneuvering 30 tons of steel, tightly berthed between two million-dollar fiberglass Hinckley yachts.

But Captain Bob's teaching manner took the panic out of

my gut. You know the answers, he seemed to imply, and we're not in a hurry. Good seamanship isn't the thoughtless instinct that salty dogs make it seem to be. It's the good habit of always asking yourself the right questions in the right order and answering them thoughtfully. I suppose that over time, as that becomes second nature, it starts to look effortless.

In just a few hours, Captain Bob gave me enough practical information and experience to make me feel I could handle my boat, which was about 10 feet longer and 10 feet higher and 20 tons heavier than any of the boats I'd handled at Chapman. He taught me not to be afraid of her, to maneuver her with big, short bursts of power, and above all to watch my momentum and put her in neutral well before I approached a dock. The first two days we tried this, I heard him but never quite remembered everything when it was time to put his advice into action. On our last trip out, though, I got it. Things just clicked. Getting my boat off the dock and feeling like I could run her were huge morale boosters. *This* was why I had come to Chapman, and though all the education I had helped prepare me, it was only Captain Bob who gave me the confidence and practical skills to competently captain the *Bossanova*.

Back at Chapman, my classmates were sweating out their final hours before the exams as I prepared for my long trip north. My observations of the first day were true—we were a motley crew—but it didn't take long for us to form a rock-hard bond built on a kind of mutinous camaraderie. I may have thought we'd have little in common, but I had overlooked the enormous fact that we were all voluntarily sitting in a classroom because we wanted to learn about seamanship. A nine-week commitment of more than forty hours a week and over $6,000 is not something anyone takes on whimsically. And spending more than nine hours a day, five days a week, with the same people, in a stressful situation where you feel allied against your captors, is an extremely bonding experience, as I'm sure studies of other hostage situations have confirmed.

It was this camaraderie more than anything else that had started tipping the balance of the Chapman experience from awful toward good.

Several weeks into our coursework, we had a midterm exam in seamanship, one of our easier classes. It covered the basic principals of boating: safety requirements, anchoring and towing techniques, and so forth. So imagine my surprise when I flunked the midterm. No, really, surprise is too mild a word. I was flabbergasted, appalled and, I'm ashamed to say, terrified. The only tests I had ever failed were occasional math tests. And not only had I studied for seamanship, but I felt fairly confident that I'd do well.

Snuggled in among the difficult questions on points-of-sail that I had spent hours (as a nonsailor) memorizing were three questions on different classes of fire extinguishers. They had seemed relatively unworthy of my attention among so much meatier fare, but they did me in. I discovered the hard way that there was no such thing as relative worth where the Coast Guard was concerned. What was worse, I was the only one who had failed, and I was now in a panic. If I couldn't pass my easiest midterm, how would I pass my other classes?

My despair over the midterm result was exacerbated by the fact that I didn't think I could study any harder than I already did. I was exhausted and anxious. But that weekend, I gave up my precious and much-needed time off to study some more. I knew that somehow I was going to have to do better, and looming before me were much tougher midterms in marine weather and chart navigation.

So I gloomily parked myself in an empty classroom on a gorgeous Saturday, and within an hour, about eight of my classmates had trickled in to join me. I knew they didn't need the extra studying half as much as I did, but they all insisted they did, and we worked together all day—reviewing everything, quizzing each other, helping each other out when one of us forgot something.

This team spirit was the most remarkable thing about my time at Chapman. As different as we all were, we felt connected to each other by our common experience. There was an unwillingness to see anybody fail, as though all of our fates were interconnected. For instance, Collin, a South African guy who had grown up around engines, spent hours reexplaining our marine engines classes to us. Without his selfless hours of brilliant tutoring, I'm positive that at least two of us would have failed our engines midterm. Instead, we got grades in the high nineties—slightly better than his, proving once again that no good deed goes unpunished.

A sense of humor was key to our survival, and we quickly developed an arsenal of inside jokes, nicknames for the teachers and each other and routines to let off steam. Most of us ate a quick lunch together almost every day at a tiny place around the corner called George's. Every few weeks someone would host a party where we'd all drink too much and talk trash. But our most touching routine developed organically, after every exam.

The Chapman dormitories, a disgraceful conglomerate of small, rundown apartments across the parking lot from the classrooms, was where about half the student body stayed— mostly the younger guys, who didn't mind the dark, dingy units that seemed to be furnished with items that had seen better days at an SRO: tattered plaid Barcaloungers, velour loveseats, paintings from the Starving Artist liquidation sale, plastic cups and glasses. The first few people to finish an exam invariably went to buy beer and brought it back with copious amounts of ice and sometimes a bottle of rum. They'd kick back on one unit's screened-in patio and hail exiting students as they trick-led out of the exam. Almost everyone came to these impromptu get-togethers. We'd compare results, commiserate and cele-brate putting one more hurdle behind us. You can't really imag-ine how oddly meaningful this tradition was unless you can picture Roger, a silver-haired, suave and successful executive knocking back a couple of Bud Lights with Mike, the pony-tailed long-distance trucker. Or Chai, a ruddy-faced, slightly insane Alaskan fisherman clinking cups with Susan, the blond pony-tailed documentary filmmaker. Or me, a staunch liberal with an "alternative lifestyle," raising a glass with John, a die-hard Republican who drove a Cadillac and referred to women as broads. ("No offense, Mare," he'd always add.)

John was usually the first out of the exam room. Short and slightly paunchy, he had a booming voice, a big heart and a broad Chicago accent. His gods were the Cubs, Auggie Busch and Bud Light—not necessarily in that order. He was only 33 but was escaping a successful career as a strip-mall developer that had caused him anxiety attacks and high blood pressure. John wasn't sure what he wanted to do in his new life, but he thought getting a captain's license was a good start.

John and I happened to be assigned to the same boat handling team, and it was in this milieu, as we chugged up the murky Manatee Pocket toward a view of the Atlantic and the freedom that was just beyond our reach, that we developed a grudging liking for each other. Despite the almost countless differences between us, we had one huge, unspoken thing in common: a love of being out on the water that bordered on narcotic addiction. Like me, John would often stretch out on the bow with his face tilted up to the sun, smiling blissfully. A former Formula speedboat owner, he had spent weekends zooming around Lake Michigan with his pals and had already mastered a lot of the rudimentary boat-handling skills that I was just learning. And though a smart Republican is an oxymoronic concept to me, John was also a very intelligent guy who quietly did well at every task the course demanded of him.

But his academic success was the only quiet thing about John. He talked a little too loudly, and this tendency became an unstoppable and amplified habit after he had a few drinks. John was supremely generous and outgoing, always insistent on picking up the tab when out with large groups, and as he drank more, he often became sentimental. ("Mare, you're the best," he'd scream, wobbling slightly on his feet. "No, no.

Listen to me. I mean it. You're all right.") More than once, we were out with John, and some overly aggressive lout would get sick of hearing John yelling and try to pick a fight. John would apologize, offer to buy the guy a drink, the guy would refuse, John would get pissed off and turn to us and say in what he probably thought was a whisper but was actually very loud, "Asshole." Then we'd have to step in and convince the belligerent fellow not to clock John, and drag him from the bar while he shouted that it wasn't his fault—it was just the way he *talked*.

Whether we were in a bar, studying for an exam or complaining en masse about the lack of modern electronics in the lab, we increasingly acted as a family. Sure, we were a very weird, diverse and sometimes dysfunctional family, but the level of care we had for each other, despite our different backgrounds and lifestyles, was truly amazing.

Near the end of our course, before the three weeks of prep for the Coast Guard exam that I had not enrolled in, we had an all-day class in emergency medicine at sea. Our teacher was a part-time EMT volunteer in his thirties, a Chapman graduate who now ran a yacht maintenance company. He was sunburned, crew-cut and walleyed. He had on new, bright white leather sneakers, jeans and a golf shirt. He had the look of a second-string high school jock who's gone soft, and his midriff was vying, unsuccessfully, for room against the beeper on his belt. He launched the class by letting us know how stupid he thought his clients were and how rich they had made him. Charming. And it went downhill from there.

He spent most of the day telling us marine emergency

horror stories. He had seen an 18-year-old dockhand get his arm ripped off when it tangled with a line as a yacht pulled away. We were treated to a vivid description of the severed limb hanging from the rope while the young man stood staring in shock at the geyser of blood shooting from his stump. Mr. Compassion had also seen a man rush into an engine room right before a steam explosion. When he emerged, his skin was hanging from his body like a loose suit.

What was ugly about our teacher was not so much that he enjoyed telling us these stories—which he obviously did—but that he seemed to relish the actual catastrophes. He also displayed a shocking callousness toward human life—advising us, for example, not to give mouth-to-mouth to the homeless, because they might have infectious diseases. He also said that he would refuse to resuscitate someone that might throw up on him—unless it was a family member, of course. He delivered his horror stories and advice with such gruesome relish that it was hard to learn anything from the lessons except the depth of his dispassion.

Unfortunately, real life provided us with a much more valuable—and tragic—lesson while we were still in school. Carol and her husband, Mike, who were buying a fancy trawler from a nearby company, went to a cocktail party their boat maker sponsored one evening. The next day, they told me about another couple they'd met there and really enjoyed. Like Carol and Mike, this couple had retired early and had just taken delivery of their luxurious 47-foot PassageMaker. They were about to set off on the first of many journeys on their beautiful new boat, and they found plenty to talk about with Carol and Mike, who were making similar plans.

Just four days later, we heard that the wife—who was only in her midfifties—had died in a tragic accident. As the couple prepared to leave the cove, their anchor line tangled with a nearby sailboat's mooring line. The sailboat was caught and pulled toward the trawler. The wife, who was on deck while her husband manned the helm, ran around the deck to fend off the sailboat and was impaled against her yacht by the sailboat's bowsprit. She was instantly crushed to death.

Probably the most common and deadly of nautical accidents occurs while fending off another boat or a dock. The desire to protect your boat from bumping into someone else's vessel or a big wooden piling is completely instinctual. Unfortunately, so is the tendency to use your hands or body weight as a force to push the object and your boat apart. Even a little bit of thought would suggest this is a lousy idea, but most people aren't thinking in this sort of situation. Chapman instructors were relentless in teaching us *never* to use any part of our bodies to fend off docks or other boats, and they definitely squelched this gut reaction in most of us.

This tragedy had more of an impact on me than anything the gruesome EMT guy had thrown at us. It highlighted how very quickly and irreversibly smart people do stupid things. The biggest danger of boating isn't necessarily the fickle weather or the merciless sea but the simple human error that leads to loss of limbs or loss of life. My boat was big, and I knew very, very little about all the things that could go wrong. For several days, I was deeply afraid of all the ways I could screw up, but I reminded myself that people die in car accidents every day, get hit by buses while crossing intersections and croak from eating undercooked hamburgers. I would be as careful as I could, but there was no point making myself sick about it.

By the time graduation rolled around, I had recovered academically. I missed a final exam on boat systems and took a zero instead of making it up: it dragged me down to a respectable 87, which was good enough for me.

Now it was time to put myself to the test. The *Bossanova* and I were going to leave Florida and head northeast. While I had initially planned on making the trip alone, I began to see the ridiculousness of this idea. It was foolish enough that my first trip off the dock for more than a few hours was going to be a trip up the entire East Coast. At my boat's cruising rate, running alone, it would be at least a four-week trip. Standing at the helm for eight hours a day, seven days a week...well, it didn't seem like the smartest idea, though I knew it could be done. Reluctantly (because I hate asking for help), I looked for someone among my classmates who'd be willing to make such a long journey without pay. There was only one classmate who was crazy enough to volunteer, and that was John. "What the hell, Mare," he said. "I've got nothing better to do. And I need the sea time. It'll be fun!"

And so, the odd couple were going to make a three-week, 1,000-mile-plus voyage, with only each other for companionship. Oh, dear. Our classmates started taking bets on who'd be murdered first.

CHAPTER FOUR

The sea hates a coward.
—EUGENE O'NEILL

June 23, 2004, was a beautiful day, on its way to being another scorcher, but it was still too early to be hot. I was
hanging over the railings of the boat with a can of navy blue
paint, writing BOSSANOVA on the gray, sun-bleached hull. John
was doing some final errands. We were getting underway
today, and I had waited until the last minute to put the new
name on the bow—though I had managed to letter the stern
the night before. My original plan had been to repaint the
entire boat soon after I closed on it, but the unexpected
bottom job at the survey had cleaned out my coffers. The
truth is, I didn't even have enough money to buy vinyl transfer letters, let alone have a professional do the fancy job the
boat deserved. I'm not sure why I had thought waiting would
solve anything—it's not like I was planning on getting any
richer soon—but I guess I dreaded doing what I knew was

going to be not such a great job. I wanted to narrow the window of embarrassment for the *Bossanova*.

I had also procrastinated because changing the name of a boat is considered extremely unlucky. A couple of my close friends had begged me to keep *Shady Lady*. Their rationale was that it was so *not* me that it was hysterical. They later confessed to a darker desire to order me a Members Only jacket with *Shady Lady* embroidered on it. Very funny—and it sums up exactly why I had to go with a new name. I remember sitting in the Hinckley office and hearing somebody bark out over their handheld VHF, "Yeah, I've got Mary here from *Shady Lady*"—every guy in the room turned a hopeful, salacious glance my way only to find not a wanton hussy but a cringing tomboy—and at that moment I knew that the old name had to go. Besides, she just felt like a *Bossanova* to me.

I'm the kind of girl who will cross a street to avoid walking under a ladder and who has practically driven into a ditch to avoid the path of a black cat. It's not so much that I believe these superstitions—I just don't see the point of tempting fate. With this in mind, I exhaustively researched boat-renaming ceremonies designed to take the bad juju out of the occasion. The Internet offered up countless rituals claiming to protect you from your own personal Poseidon adventure. Some demanded that you perform the renaming with the boat in the water, others insisted it be in dry dock. Some called for champagne offerings, and others for red wine. One stipulated sailing backward for 200 yards, and another predicted bad luck until you'd run aground three times. It was impossible to determine which ceremony was best, and in the end I just borrowed from a couple and created my own. I figured that if I respectfully asked Neptune/Poseidon for a blessing on my boat with a pure heart, the ritual itself could be safely improvised.

I did embrace the widely held notion that renaming a boat is a two-tiered process. It just made good sense. Lore has it that Neptune/Poseidon keeps a ledger with the name of every boat recorded therein, and the first thing you have to do is purge the old name from his ledger and his memory. It also seriously pisses off the Big Guy of the Deep if you bring anything aboard with the boat's new name on it until you've eradicated the old. In this respect, I inadvertently screwed up—I did have a couple of e-mails on my computer that referred to the boat by its new name. I also had my running papers aboard, including my Coast Guard documentation with the new name of the vessel on it. But I guessed that the deities of the deep would be willing to let these tiny transgressions slide. After all, they're probably not computer-savvy enough to tap into my hard drive and I did have a legal obligation to keep the paperwork handy.

I began my ritual by first removing every trace of the vessel's old name. This was harder than you might think. It meant not only peeling the old lettering off the bow and stern, but painting over the ring buoys, removing all of the old documentation, maintenance records and ship log. (I mailed them to a friend to keep for me.) Finally, as I took a happy look around the salon, satisfied that I had eradicated every reference to *Shady Lady*, I saw the previous owner's cruising card pinned to the bulletin board. His name and contact information were superimposed on a photo of the boat, and when I squinted, I could see *Shady Lady* written across the bow. I left the card but blackened out the lettering on the boat photo with a Sharpie, hugely relieved that this crucial detail had not slipped by me.

The next thing I did was gather the dogs in the salon and give a little speech, praising *Shady Lady* for her years of faithful service to Mel and asking Poseidon to erase her esteemed name from his ledger. Then I thanked him and raised an icy margarita in his honor. I also gave each dog a Snausage. We all savored the moment.

Then, I asked Poseidon to record the new name, *Bossanova*, on his ledger. I asked him for his blessing on this name and implored him to keep the boat and all her passengers safe. I closed by hoping he'd help me be a captain worthy of my vessel and of his constant protection. I raised my glass to him again and then I turned and toasted all four directions of the wind, just to be safe. The dogs looked at me expectantly (which is pretty much how they always look), so I obliged with another treat. They looked at me expectantly three more times. Smart *and* greedy!

Later that day, after I had painted the name on the stern, I went forward with another margarita, splashed it liberally across the anchor locker and bow and over the side into the water. Then I officially pronounced the vessel *Bossanova* and drank the remainder of the cocktail. I knew the margarita was an unorthodox alternative to the traditional blessing with good champagne but, in my mind, the quality of my offering was what mattered. And I make a mean margarita.

All in all, I was happy with how the dogs and I had welcomed *Bossanova* into the world, but I was decidedly not happy with my hand-painted lettering job.

I am not a vain girl. Sometimes I go days without even glancing in a mirror, and then when I finally do, I think nothing more than "Yup, that's me, all right." A friend once described my rumpled sartorial style as "home from St. Andrew's for the weekend," and I noticed recently that all my childhood photos showed me wearing khakis and navy blue sweaters, an urge I still have to fight daily. But I felt real pain whenever I failed to keep *Bossanova* looking good. I thought my boat was gorgeous. I was proud of her and I felt a duty to keep her looking her best.

The new name looked awful. A couple of guys stopped to tease me about it on their way up the dock to *Storm-Along*, a beautiful motor yacht with lots of varnished wood and a distinctive inky green hull with matching upholstery. This vessel, which they were busily making repairs to, was not much bigger than mine but was probably worth well over $1 million. The nice thing was that—and this happened *all* the time—these same guys had stopped by the day before to say how much they liked my boat and to ask me all about her. So, I took the ribbing good-naturedly and prayed they wouldn't see the stern. I felt deeply ashamed, like I was sending my kid to school in filthy clothes.

Honestly, it really was a minor disaster. After I had removed the old vinyl letters that spelled out SHADY LADY and painted over the area with a rectangle of fresh gray paint, I thoughtfully eyeballed the shadow of the previous name's outline, whose glucy ghost still peeked through, ever so faintly. I figured that if I kept the same letter spacing, it would look pretty good. A nice theory, but when I stepped back to have a look, I had somehow crammed all the letters together on one end. It was completely lopsided and looked like someone with significant learning disabilities had taken a stab at it—with her foot. My friend Julie later coined a name for my hand-painted font: Retardica Bold. I vowed to repaint it while underway—it was too embarrassing to imagine arriving in Sag Harbor like this.

John came down the dock, brandishing an air filter triumphantly. The previous owner had run *Shady Lady* without one for more than ten years and never had a problem. I had absolute faith in Mel's advice and I also subscribed to the motto If it ain't broke don't fix it. But there was an obvious basket attached to the engine that cried out, in its naked emptiness, for a filter. Clearly the manufacturer had intended one. Chapman had also drilled into us the importance of keeping dirt and foreign bodies out of the engine. With two broken-coat Jack Russells aboard, I had dog-fur tumbleweeds if I didn't sweep on the half hour. So I decided to err on the side of caution.

The filter had been easy enough to find at NAPA, but installing it was a bear. The bolts on the holder had ossified from disuse and John had a tough time getting them off. Once he had the filter installed, he couldn't close the bracket again. But he used a few small pieces of wire to keep it in place. We decided that a jury-rigged filter was better than no filter at all.

At this point, I was pretty determined that nothing was going to keep us from getting underway. The week before, I'd been trying to do some small maintenance to prepare for our trip. The raw water strainers, which supply cooling water to circulate around the super-hot diesel when it's running, needed to be cleaned. They were black with gunk from the boat's trip across the Okeechobee Waterway to Stuart and its subsequent lingering soak in the murky Manatee Pocket.

Wrestling the tops off the overtightened strainer lids had caused the handle to crack, and some temporary leaking, which disappeared when the system was pressurized, alarmed me enough to order a replacement. In the end, I just kept the new one as a spare because the old one continued to function. But waiting for the part had delayed our departure by two days. Even now I suspected there were a few things we'd overlooked or forgotten in our preparations, but I also felt that we were as ready as we'd ever be. No doubt there is always one more thing you should do, but we weren't going to. It was time to head out.

And so, at 1130 hours, as John cast off our lines, I backed us out of the slip and powered us to starboard in a tight circle. I headed the *Bossanova* up the Manatee Pocket and toward the Intracoastal Waterway, and as we passed the channel into the Chapman docks, I sounded the horn.

It was John's first time out on the *Bossanova*, and only the third time I had taken the boat off the dock myself. I felt elation, giddy excitement. But in the back of my mind hovered a subtle, fluttering sense of irresponsibility, swatted away by sheer determination. I didn't really have enough experience to be making a journey like this, and I knew a million things could go wrong: terrible weather, disastrous mechanical problems, contaminated fuel, running aground. I wasn't really prepared to handle any of these scenarios. But I focused on the here and now: *Stay in the channel, watch your rpms, check the oil pressure, watch for traffic astern, monitor Channel 16....What a beautiful day. Oh my god, we're underway and I am the captain of this beautiful little ship. It's a dream come true.*

Halfway up the pocket, I planned to stop and fuel up for the voyage. This would be the very first time I docked my boat solo, and even though I'd merely be coming alongside an open bulkhead, I held my breath. I didn't want to mess up—it would seem like a bad omen. Adding to the pressure was a Coast Guard patrol boat, already at the dock for refueling. A posse of young Coasties in uniform eyed me skeptically as I made a 180-degree turn and brought us alongside the fuel dock. Perfect. I had to remind myself that seeming too happy about it would be very uncool and totally blow my cover as an ancient mariner.

Three hundred and sixty gallons of diesel fuel and $600 later, John and I headed back up the pocket and into the Intracoastal. We anchored that day at 1910 hours, at Pepper Point, otherwise known as Mile 935. My logbook shows readings from the engine gauges throughout the day and notes the times at which we passed the Fort Pierce Inlet and Vero Beach. After a few days, this kind of overly meticulous logging fell by the wayside and we recorded major events, weather conditions and anchorage spots. Of course, I still

monitored the engine gauges with great regularity, but everything continued to look stable and I stopped writing it down
over and over, like a big nerd.

Our first night's anchorage was a deep spot outside the
ICW channel. We knocked off around 1830 that evening, and
once we had securely anchored and shut down the engine, we
sat on the stern with a couple of cold margaritas and toasted a
fantastic first day. The air was warm and held that magnificent pinky golden hue particular to summer evenings. John
and I felt not only a deep sense of contentment but also
relief.

The scenery that day had been gorgeous and ever-changing.
I kept a couple of guides handy in the pilothouse, and from
time to time, I tried to identify the beautiful vegetation and
wildlife we encountered. We passed by mangrove islands
shaded by oak trees, red cedars and cabbage palms. I spotted
what I thought were prickly pear cacti, covered with beautiful
yellow flowers. Sea oxeye daisy, mangrove, saltwort and spar-
tina grass lined the banks. Wading birds, including shorebirds,
ospreys, cormorants, brown pelicans, roseate spoonbills and
wood storks, watched us slowly passing, rarely perturbed
enough to fly away. There were dozens of dolphins, a couple of
manatees, and plenty of people fishing from the banks. It felt
sleepy and peaceful as we chugged along through our first day
without even a hiccup. After all those months of sitting in a
classroom, we were finally experiencing what had drawn us to
school in the first place. We were really *doing* it! And we seemed
to be on top of everything.

Of course, we had no way of knowing that very few days

would be as uneventfully successful as our first. Even the very next morning, shortly after we left our anchorage at 0715 hours, we started noticing some problems with the VHF radio that would plague us through most of our trip. I suspected it was battery-related. The diesel engine was fitted with an alternator that charged the batteries and allowed the essential electronics to run off their stored energy while underway. I had wondered if my batteries needed replacement even before we left the dock, but a marine electrician had tested them, swapped out some old wires for new ones and declared them absolutely fine. Now, though, our first radio check of the morning showed a pathetically weak signal. But as the day wore on and the alternator continued sending power to the batteries, our signal seemed to recover. Maybe our radio was fine after all.

The second day was memorable mostly for a far-off view of Cape Canaveral that was thrilling in the early dawn light. All we could see was the profile of the launch pad, silhouetted against the pink morning sky. It looked majestic, though distant. At eleven o'clock, we could make out a little more detail in the midday sun but by one o'clock, we had run out of enthusiasm for it. Around three in the afternoon we started to actively avert our gaze, looking at anything but the Space Center. And at five o'clock, when Cape Canaveral seemed as far away as it had at seven-thirty that morning, we became slightly unhinged and took turns hurling insults at its taunting immobility. Nothing deflates a sense of accomplishment like watching the same landmark go by…all…day…long. The contentment of the first day was already being invaded by a teeny, tiny, vaguely itchy realization that it was going to be a long trip at this rate.

In the late afternoon of day two, we stopped at the Titusville Municipal Marina for ice and then motored on for another few hours. It was earlier in the evening than we had wanted to stop, but the route ahead looked like a slow and twisty one through an area that some dimwit had enticingly named Mosquito Lagoon. I later learned that this spot had been rotary-ditched back in the 1950s in an effort to control the mosquito population. From Jupiter Inlet in the south to Ponce de Leon, which was still north of us, this was all part of a 156-mile-long estuary called the Indian River Lagoon, the "Redfish Capital of the World." It claims to be the U.S.'s most diverse estuary, with over 400 species of fish, 250 species of mollusks and 475 species of shrimp. At Mile 860, though there was nary a mosquito in sight, we were being bombarded by the big black horseflies we'd been warned to expect in North Carolina—obviously, ours were precocious. Maybe they'd killed off the mosquitoes. We pulled outside the channel and into an area where the river widened, in an attempt to outsmart the little buggers. It worked.

When we went to bed we shut everything down, and in the morning, it seemed as though the batteries had held most of their charge overnight. The VHF radio was again working beautifully, which was imperative if we were even going to think about venturing into the Atlantic. The VHF is often your only (and always your most important) means of communication aboard a ship. It's the source of vital weather updates, special notices of military maneuvers, and warnings for unscheduled bridge closings or dangerous objects afloat. It's also your lifeline to help. In case of a mechanical failure, a fire, a medical emergency—anything that might cause you to summon immediate help or abandon ship—the VHF alerts not just the Coast Guard but everyone aboard a boat within a 20-mile radius of your situation. It would be unthinkable to go offshore without it. Even hearing it fade and die as we ran the Intracoastal the day before had made me anxious. Its miraculous recovery was excellent news.

A few hours later in New Smyrna, after our slow meander through the twisty area we'd avoided the night before, the landscape had completely changed. We were motoring through a suburban neighborhood of ranch houses with screened-in pool rooms and condos with views of the Intracoastal. Ahead of us, a bedraggled sailboat putt-putted along at a tortoise's pace. Its deck was cluttered with jerry cans and drying clothes; a battered inflatable dinghy bobbed along in its wake; its hull was painted in a patchwork of faded colors. This boat looked like it had dragged itself halfway around the world and was now officially exhausted. After a while, at a very wide point in the channel, I signaled that I was going to pass. We were happily waved ahead, but within a few seconds we felt a sickening *thump*.

Mariners always say that the only people who haven't run aground in the ICW are liars.

Still, it wasn't until I had powered us off the shoal, fifteen minutes later, that I was able to joke that I was glad to get that little tradition out of the way. Of course, I could afford to laugh. At 30 tons, the *Bossanova*'s steel hull was built like a tank and her unusual box keel meant the propeller shaft exited the boat in a straight line, making it less vulnerable than a typical (angled) shaft. When I had gunned the engine enough to kick up great balloons of sand in the water, we finally floated free and were none the worse for wear. I was now even more impressed with my sturdy little ship and her excellent design. Later that morning, John would also very briefly touch ground, twice in several minutes, as we negotiated a portion of the channel that had become very shallow. It occurred to me that, technically, since we'd now run aground three times, any curse the name change might have brought should be history. I worried about the technicalities, though: did touching ground count as running aground? I suspected we had to get hung up to get credit for the incident. That would happen soon enough.

Shoaling in the ICW is a hot-button issue among boaters in recent years. This federally maintained system of waterways that's most traveled between Norfolk, Virginia, and Miami, Florida, is made up of natural rivers and estuaries connected by man-made channels. Periodic dredging is all it takes to maintain this boater's highway that parallels the Atlantic Ocean but is protected behind the coast. The authorized depth of the Atlantic ICW, meaning its minimum depth at low tide, is 12 feet anywhere along its course between Norfolk and Fort Pierce, Florida. (From Fort Pierce to Miami, the authorized depth is 10 feet.) Most of the ICW falls below 12 feet at low tide, though, with many, many areas dropping as low as 5 feet.

In recent years, federal funding to maintain the waterway has been cut and dredging has fallen behind schedule. Some people suspect that since most commercial cargo is now moved around the country by trucks, planes and trains, government enthusiasm for maintaining the ICW has ebbed. However, an entire economy exists along the waterway, and is as dependent on the seasonal migrations of boaters for their livelihoods as these "snowbirds" are on the safety of the ICW.

For most recreational boaters, the ICW is the only way to go. If time's not an issue, the ditch is scenic and relatively safe. But after only two days of it, we were getting antsy. Time was an issue for us. John had already scheduled a flight from New York to Chicago, and I had some much-needed work waiting for me. Even though we were only a few days into the trip, we could see from the charts of the winding ICW that we were going to be underway for a very long time, much longer than we had planned. And to be honest, John and I both had a hankering for the real sea, not this sleepy two-lane highway.

As we approached the Ponce de Leon Inlet, we made a couple of quick calculations and decided that we would have plenty of time to make St. Augustine by nightfall if we went offshore.

We double-checked our VHF radio and the marine weather forecast. Both were good. We had one more look at our guide to the Atlantic inlets. Alarmingly, Ponce de Leon Inlet was listed as one of the most dangerous on the East Coast. But further reading seemed to suggest that many other inlets shared this dubious distinction. While I was at the helm, John chatted on the VHF with a sport fisherman who passed us. He gave us some tips on where to look out for unmarked shoaling, but told us it was an otherwise very manageable passage.

Our Chapman experience had included one afternoon when we took a boat out to the St. Lucie Inlet. I'd have to say that running this inlet several times was the most exciting thing that happened at Chapman, with the exception of acing my midterm in chart navigation. Transiting an inlet safely is something that takes skill and an understanding of the forces of water, wind and tide. For instance, a strong onshore wind with an ebb tide is a combination likely to produce dangerous waves.

Passing through an inlet on your way to the sea has the advantage of showing you the face of the waves. In other words, you can see how rough the conditions are and decide to turn around if it looks too difficult. There's no shame in this option—at least a half-dozen people die every year just trying to run the inlets of southeast Florida.

Passing through an inlet from the ocean can be even trickier. Coming from the sea, you do not have the advantage of seeing the wave's face; you can only see its back—the swell rolling toward the shore. This is when most accidents happen. One danger is that your bow will slow and your stern will get kicked out to the side, causing you to be beam-to and broadsided by the surf. Very perilous and not a pretty thing to see— a fabulous way to get swamped and sink. But even worse than that is being caught on the downward slope of the wrong wave. If the angle is steep enough and the wave period short enough, your stern can get flipped right over your bow by the following sea. There is a word for this, "pitchpoling," and its very mention sends a shiver down any mariner's spine.

The correct way to enter a rough inlet is to ride the back of a wave—not the crest!—but this requires good timing and practice to keep the throttle at just the right speed. When you can do it perfectly, it's a subtle thrill, very like bodysurfing, to feel nature carrying you safely through the rough and pushing you into calmer waters.

Buoyed by our Chapman experience and craving the wide-open ocean, John and I agreed we were ready to brave the Ponce de Leon Inlet and see what was out there.

Using the tips the passing boat had given us, we knew exactly which markers to stay close to and where to look to port for another channel to take us out. When we reached the inlet, the surf was low and rolling and the water was flat and bright. We motored through easily and it was wonderfully anticlimactic. Days without drama are exactly what a good mariner craves, and they happen less often than you'd wish, as we'd soon find out.

Once we were on the outside, I stood at the helm and steered us out. As the shore and its signs of civilization grew smaller, the sea flashed its amphibious jewels and fluidly turned a dozen shades of blue, green, gray. I was giddy but apprehensive. This was the big time—my little ship's first foray into the Atlantic—and I prayed that nothing would go wrong: no mechanical failures, no bad weather, no hidden navigational hazards. After an hour, when I had corrected our course to run parallel to the shore and switched on the autopilot for the first time, I slowly let out my breath and relaxed. I listened to the Coast Guard on the VHF, checked my instruments, watched as Daytona Beach, Ormond-by-the-Sea and Flagler Beach slipped by in the distance. And I thought my heart would burst from the ecstatic sensation that I was exactly where I wanted to be. I had been right: this *was* the life for me.

CHAPTER FIVE

Black sea, deep sea, you dangle
Beneath my bliss like a dreadful gamble.
—JOHN UPDIKE

Once we were on the outside, it was clear that there was no going back. Looking out toward Portugal, it was hard to discern the sea's horizon from the sky. I felt like I'd been released from a cage, and the sense of freedom was so intoxicating that I had to resist the urge to stand at the bow, fling my arms out and shout, "I'm the king of the world!"

But there was no point alarming John this early in our trip, and it was probably better not to invoke the *Titanic* right now.

All other considerations aside, if we continued to run up the coast, instead of through the twisty-turny ditch, as the ICW is known, we would save ourselves a lot of time. So John and I quickly agreed to leave its pokey charms without a second thought, gladly trading the murky rivers for the cornflower blue of the Florida Atlantic in sunshine; the cypress, pine and moss-draped oaks for frolicking dolphins and giant sea turtles; the ambling pace for, well, a slightly less ambling pace.

In the Atlantic, you don't have to worry about shoaling channels, oncoming traffic or scheduled bridge openings. You can set the autopilot, keep an eye on the course and instruments, and relax a little. In return, you give up the utter safety of the ICW, the free anchorages in quiet spots, the interesting sights and sounds of life on either bank as you pass through. But for us, there was simply no question that the Atlantic route was well worth its slightly riskier ride.

Everything went smoothly for us on our first day offshore. John and I took turns at the helm. The diesel was very loud when the engine was running and the bridge was the quietest spot on the ship, so we kept each other company even when it wasn't our watch. We were both ecstatic about the sights and sensations of being offshore: the constantly changing colors of the water, the dolphins and giant sea turtles and large schools of fish we passed through. When we weren't chatting and I was at the helm, John read a book—some conservative pundit's attack on liberals. I ignored this. Around 1:00 p.m., I made us turkey sandwiches with pickles and potato chips on paper plates and we ate in the pilothouse. We tried tuning in a radio station, but the reception wasn't good, and anyway, we were both afraid to run down our iffy batteries by wasting power.

Our first afternoon was relaxed and happy, so when we made our planned destination, St. Augustine, earlier than we'd anticipated, we were tempted to just keep going. We still had a few hours of daylight left, and after we checked the chart, we decided we could make it to Jacksonville Beach by nightfall.

Forty-five minutes from the entrance to the Jacksonville Beach channel, I caught the Coast Guard broadcasting on Channel 16—saying something about severe weather, I thought. Our skies looked fine, but John and I fell silent while I switched the VHF up to Channel 22 for the full announcement.

Attention all stations, attention all stations. This is the United States Coast Guard, Mayport Group, United States Coast Guard, Mayport Group. At approximately 1800 hours the United States Coast Guard was notified of a storm moving toward Jacksonville Beach, in the vicinity of 30 degrees 20 minutes north and 81 degrees 36 minutes west. The storm is moving southwest with winds of up to 55 miles per hour and will be accompanied by severe lightning and thunder. All mariners are urged to seek safe harbor immediately. Again, all mariners are urged to seek safe harbor immediately. This is the United States Coast Guard, Mayport Group, Out.

Out. The word echoed in my mind with ominous finality. "Out" is radiospeak for "signing off," but I thought I detected something like pity in its curtness. It sounded more like a dubiously offered "Good luck...because you're going to need it."

I had a funny feeling in the pit of my stomach: half dread, half excitement. For those in the vicinity aboard something sporty like a little center console with twin 250 horsepower engines, this warning was useful. All they needed to do now was point their bow toward shore and gun it. But for us, the

Coast Guard's announcement was nothing more than a bad review of a show where we had already taken our expensive and highly visible front-row seats. There was no chance of escaping now.

John and I put our heads together over the chart. The only way we could possibly find shelter was to make it to the closest inlet: Jacksonville Beach. But as the skies darkened to the ominous green and black of an old bruise, it became clear we couldn't beat this monster front. It was moving at us—fast—from the *direction* of Jacksonville Beach, and we decided there was no point going right into the storm. Instead, we turned 180 degrees and ran away from it. Of course, when your boat averages about 7.5 knots (approximately 8½ miles) per hour, "ran" is just a figure of speech. What we were doing was more like shuffling.

A storm at sea is something to see. First, imagine being at the beach and watching a big storm from behind a plate glass window—the vastness of the ocean, the change in the waves as they grow and crash against the sand, the way the skies become almost black beyond the surf. The rain lashes against the window, streaming down the glass in a blur that makes it hard to see exactly what's happening out there. A storm at the seaside gives nature a wonderful stage for its high drama. But the overall sensation you feel as you watch is one of coziness. You're safe inside. You don't have to go out. Start a fire. Have a cup of tea. Enjoy the show.

Now imagine that you are surrounded by that dark sea, bobbing in the middle of it. Instead of looking out a big picture window, you are looking out the blurry windows of your pilothouse as your little ship rocks from side to side and climbs steeply up one wave and down another. Off in the distance you see land and it looks...like heaven.

We did not make tea. One of us stayed at the helm while the other ran below to check that all our portholes were tight and all loose objects were stowed. The dogs took up a cowering position in the corner of the pilothouse bench. We put on our foul-weather jackets and felt as prepared as we could be.

Bright white flashes of lightning were now shattering the charcoal sky behind us, and we were feeling pretty smart for evading the brunt of it. We decided to move a little farther off the coast and out, then circle back up toward Jacksonville Beach. We changed course, and half an hour later, it looked like our tactics had worked: all but the fringes of this brief, violent front just passed us by. Of course, it was our first storm and we had nothing to compare it against. Later I learned that 21,000 people ashore lost their power that night and more than 10,000 remained without electricity for another 24 hours. We were right to feel smug.

But that sensation didn't last long. When we regained our previous position just offshore of Jacksonville Beach, it was pitch dark and the storm had left turbulent seas in its wake. While I wrestled us over 8-foot waves at close intervals, John tried locating the entrance to the channel, relative to our position. We hadn't studied our approach in advance, assuming instead that we'd be there in daylight under good conditions, with plenty of time to strategize en route. Then the storm hit, and we focused solely on staying out of its way. Now we realized we'd made a very amateur mistake: we had failed to familiarize ourselves with the approach to an unknown harbor in advance.

According to the chart, the entrance to the channel was about two nautical miles offshore, marked by a flashing red and flashing green marker. But how far out did the jetties extend? How deep was the water approaching the entrance? In daylight, I would have felt confident about cutting into the channel without going all the way out to the first marker. But in darkness, I was afraid of what I couldn't see (since charts are not always accurate), so we headed all the way offshore to the top of the channel entrance. The *Bossanova* rode the waves like a cowboy atop a bucking bronco—up the face of one wave and steeply down the face of the next. And we couldn't see a damn thing.

To reduce glare, the only light on in the pilothouse was a red bulb, but I still had to leave the doors open and frequently step outside, with one hand on the wheel, to get a clearer view of the ocean at night. Even the dimmest light on the bridge seemed to ricochet off the glass and obscure the outside. But standing in the open air, my eyes gradually adjusted to the slight variations of blackness around us. Beneath, before, behind—that dense blackness was sea. The parts that seemed trimmed in a faintly luminous frill were the crests of waves. Above and all around us, the night sky's scattered stars drew the only discernible border between the dark waters they blanketed. Every minute or so, I stepped back out, let my eyes adjust, had a good look around and then stepped back in. When we finally headed into the very top of the channel entrance, it was 9:00 p.m. and we'd been running for twelve hours.

As we passed through the jetties, just past the mouth of the St. John's River, we sighted some secondary channel markers to port. These were a line of smaller buoys leading to Mayport Basin, the third largest naval base in the United States. The base was lit up like a massive, open-air operating room. An aircraft carrier tied by a dozen hawser lines to the dock resembled a dangerous giant, strapped down to sleep off anesthesia under merciless white lights.

Unfortunately, these lights, and all the other less spectacular shore illuminations of Jacksonville Beach, made picking out markers in the main channel nearly impossible. Now when I stepped outside the pilothouse for a look, there was no darkness to adjust to, just a different kind of light. In its own way, it was as stressful as finding our entrance had been. We were safer in that we were not out there alone, pitching around in a tar black sea. On the other hand, we were in a major navigation area that was unfamiliar to us, we didn't know where we were going and we couldn't see anything.

We took turns—one of us at the helm, the other outside the pilothouse with binoculars and a search light, which we swung like a scythe through the blackness, carving a path up the channel, one marker at a time. When the beam weakly hit a distant buoy, we focused intently, calling out its number as soon as we could see it and checking it against the electronic chart. We were slowly groping our way forward.

Theoretically, this nighttime navigation should not have been as tricky for us as it was. We had both suffered through a very difficult (and excruciatingly dull) course at Chapman called Rules of the Road. Success depended on hours of sheer rote memorization of Coast Guard legalese covering which vessel has right-of-way in any given situation and which day shapes or sequence of lights marked which kind of vessel. The Coast Guard licensing exam required a grade of 90 percent or higher on this section of the test in order to pass, so the material was drilled into us. We were quizzed at the beginning of each and every lesson on the previous dose of mind-numbing rules, and in the end, we all squeaked by. But as real life was teaching us tonight, all the other ambient lights in a port—including highway lights, commercial signage, secondary channel markers, even passing cars—make isolating the identification lights of other marine traffic and navigation aids very tough.

Now, as I looked up the river, I saw a stack of lights moving across the horizon, parallel to us. "What's three whites over a green mean again?" I asked John, who had done an extra three weeks of extensive preparation for the Coast Guard licensing exams.

"Hmmm," he said. "Let me see. Three whites over a green...three whites over a—holy shit, that's a *long tow*." A long tow is a vessel of 200 meters or more, being towed by a tug. And that was only the *minimum* length of the barge. We were in so much shock that I didn't try to calculate its actual length, but it was gigantic as it rounded the bend and headed straight for my wee 40-foot boat. It absolutely dwarfed us. I moved over to the starboard side of the channel in order to give this leviathan plenty of space as it closed in. But about a minute later, we heard two prolonged blasts and one short blast ring out from behind us. John stepped out of the pilothouse and shouted back in "Oh my god, Cap. I cannot believe it. We've got another long tow coming up behind us."

When you're out in the ocean, with plenty of room around, you still give these giants as much leeway as possible. They throw off an enormous wake and they're very restricted in their ability to maneuver. Now I held us steady in the middle of the river as we were passed on either side by barges 600 feet long or more, passing on both sides of us and no more than 100 feet away.

There was really nothing I could do besides alter my course ever so slightly and get ready for the fallout. Nothing we studied could have prepared us for this, and I was vaguely aware that—in terms of channel traffic nightmares, at least— we were being given a scenario about as bad as we were ever liable to meet. Still, after months of stressful study in a classroom, something in me rejoiced that we were out from behind our desks and books and actually *doing it*. I wasn't half as scared as I was thrilled.

The two long tows passed us simultaneously, and for a moment we were sandwiched between their hulls. Piled high with cargo, the ships caused a quick eclipse of the stars as they passed us. The channel waters they left in their wake were confused, meeting in the middle in a high swell that rolled us gently back and forth.

And then they were gone. And we were fine.

An hour or so later, we finally found a deserted dock to tie up at for the night and almost wept with relief. It was midnight, the end of a seventeen-hour day underway, and our first one out in the ocean. My body was absolutely exhausted but my heart ached with happiness. After giving the poor boys a quick walk, I fell into bed, fully clothed.

What seemed like minutes later, when I could no longer ignore the clamor chewing at the edges of my consciousness, I opened my eyes. It was 5:15. Where was I? Engines were coughing and catching, I heard the high-pitched beep-beep-beep of a forklift in reverse and someone shouting, "Back it up, back it up." Three boats in quick succession peeled away in loud, whiney flourishes. Was that diesel I smelled or testosterone? It was still gray out, but it was definitely time to get up.

It slowly came back to me. John and I had finally turned in about four hours earlier. We'd had a storm, a rough approach to the channel, and a cheek-to-cheek dance with two long tows. What we needed was four days of sleep, not four hours.

But we had tied up near the fuel dock of the only marina we could find and, of course, today had to be the day they were hosting their annual fishing tournament. I threw on shorts and a clean button-down and took the dogs for a walk to the office so I could pay for the overnight tie-up. A plump girl with freckles, maybe 10 years old, was handling the register like a pro, and a line of sunburned guys in shorts and grimy caps hustled through as she rang up their purchases. Everyone seemed a tad morose and hurried. The marina office looked and felt like any small-town gas station–minimart, except for the assortment of lures, baits, charts, boots, cheap sunglasses, spare starter batteries and fishing equipment that lined the shelves along the walls. All the other offerings were typical of a highway convenience store, right down to the beef jerky, snack cakes and pump canisters of coffee. I skipped the array of flavored nondairy creamers and had a sip of tepid coffee-flavored water from a Styrofoam cup while I waited in line. When it was my turn, the bubbly little girl said she didn't know how much to charge me and got on the VHF to someone named Randy.

"I've got someone here wants to pay for tying up last night. What do I charge her? Yeah, that big boat by the fuel dock....Okay." The girl pulled a large ring binder from under the counter and handed me a slip of paper with questions: name of vessel, length, beam, date.

"Looks like it's going to be busy today," I offered while filling out my form. I'm a mistress of the obvious.

"Oh, jeez. You have no idea. And the worst thing is, a lot of people are pretty angry because we, like, ran out of diesel."

So, that *was* testosterone I had smelled! "Oh, dear. And you guys have a big tournament today, huh?"

"Yeah, and it's, like, a long way back to another place with fuel, so they're not too happy with us. And it's not so good for us either because lots of these people are our regular customers."

But she said all this with a cheery smile, took my paperwork, checked the binder and came up with a charge of $80. Eighty dollars for about four hours: it seemed steep, but she was sweet and I was too tired to argue.

Back at the boat, John and I agreed we should get outside again as fast as we could. He went for ice and I put the water on for real coffee. This was before my life-changing purchase of the beautifully basic stovetop Bialetti Moka espresso maker, so I was still pouring boiling water over Bustelo through the wrong-size paper filter wedged in a small red plastic funnel I'd found in the engine room. It wasn't pretty but it did the job.

We were underway by 6:00 a.m. It was dawn as we passed between the jetties and the sun flashed patches of liquid gold on the dark Atlantic. It was a beautiful sight, though the noisy escort of center consoles on either side of us, opening their throttles to full drone as soon as they left the No Wake zone, pissed all over God's grandeur.

Once we were well offshore and headed north, John and I came up with the day's game plan. This would be our routine every morning. We'd roll out a paper chart and have a look at the coastal towns up ahead. Which one was the farthest away but still reachable in daylight and close to an inlet? We'd take a guess, then plot a line or series of lines along the coast and measure the distance using a compass divider, comparing it against the chart's distance key. Then we'd calculate how many hours it would take us at cruising speed and verify that we could, indeed, make our guessed-at arrival point by evening. If we had chosen a place that was a tad too far, we'd back up and find a closer destination. If we'd been underambitious, we'd look for the next inlet up the coast and run calculations again to be sure we could make it by nightfall. Then we'd chart our first leg and put the coordinates in the GPS. The GPS would then show us a bearing and an expected time of arrival that confirmed our hand-drawn work. We'd then set the autopilot and make more coffee. This morning, we were exhausted but happy, and we spent the next few hours rehashing the previous night's incredible adventure.

After a second cup of coffee, we talked about the possibility of an overnight run to Charleston. We decided we'd go as far as we could by day and see if the weather held, and if we felt up to the challenge of navigating by night.

At noon, John donned his radio headphones and took his fishing pole to the stern of the boat to listen to a nationally syndicated sports radio show from his beloved Chicago as well as the latest in right-wing punditry. I stayed on the bridge and admired the unspoken separation of church and state that kept the crew of the motor vessel *Bossanova* fond of each other.

ONE THING YOU HAVE on an ocean voyage is time. The sun goes in and out of a cloudy sky until it tires of the chase and declares itself against a bright blue ceiling. The waves get shorter, get taller, come at you from different directions, take on different gradations of blue and green. Every degree of latitude brings a slightly different feeling—new wildlife, a crisper light, a slightly different smell. The sea's surface mirrors heaven's every nuance. Watching this and not thinking about anything at all was what I loved best.

There was something about looking at the coast from a distance: an awareness that I was not—for now—a part of life on earth, an ant on the farm that toiled back there, that stopped to pump gas and get groceries, that had dentist appointments or social obligations. I felt apart, invigorated, clean. This was my life right now—standing at the helm, checking the GPS coordinates against the chart, keeping an eye on the radar and the autopilot, stepping outside with the binoculars to determine a far-off freighter's course, switching to WX on the VHF for the weather report and turning up the volume every time I heard a Coast Guard bulletin. Just running my little ship.

But there was plenty of time, too, for thought. Mine is a restless mind, but the rhythms of life aboard seemed to quell that combustion chamber where compressed thoughts ignited worries. When I thought, I reflected idly. I didn't try to figure stuff out, resolve anything, take it apart and understand it— these were all hallmarks of my landlubber mind.

I have always been good at what I can understand. But who understands love or the reason it falters? My failure at that one thing I valued more than all others had played a big role in my decision to veer off a straight life course. And now, at sea, I found myself able to review pieces of my past without too much analysis. It was like throwing away the microscope and suddenly realizing you could see better with your naked eye.

When I was 24 I had a girlfriend named Maud. She was eight years older than I, but she was so young at heart that when she walked down 14th Street in New York City, 16-year-old homeys would come up to her and tell her she was *dope*. She wore Converse high-tops almost every day, with jeans and one of those black-and-white tweedy windbreakers with leather sleeves. She had leopard patterns carved in the hair on the back of her head and she had faux-marbleized her old-model Volvo wagon. She was a very talented writer, photographer, and chef. She also drank a little too much and was truly agoraphobic.

But what drew me to her was her almost pathological charm. She was always promising to call people or see people and then blowing them off but winning them back. She did that with me, too, and the power of her regret was always so much more endearing than her carelessness was painful.

We met when I was living in Nantucket, painting houses and trying to figure out what I really wanted to do. Maud and I developed a ritual of dinner and a movie once a week. I would ride my motorcycle to the Finast and pick up steaks and artichokes, get a bottle of red wine, and then rent a couple of movies she'd selected. I had watched very little television and seen very few movies during my childhood. Maud gave me a crash course in great films. We would drink wine and talk about art and literature and politics while she cooked. When dinner was ready, we'd sit in her darkened living room in Adirondack chairs with our plates on our laps and watch *Mildred Pierce*, followed by *Double Indemnity*. Or *I Want to Live*, followed by *Hiroshima, Mon Amour*.

After about a year, I felt discouraged by our intense yet casual relationship and by life on the island. I thought Maud didn't take me seriously because I was younger. In fact, she often pushed me away with the excuse that I shouldn't be weighed down by someone with as many problems as she had. Eventually, I gave up and moved to New York. Within a few months, Maud surprised me by following. We shared a loft on 19th Street with the landlord's cat.

Maud, who let's not forget was agoraphobic, got a job at a

sewing pattern company, where she had to ride an elevator every day and punch in and out, like a regular stiff. She was a trooper. I, meanwhile, had fallen into a highly coveted job as an editorial assistant at Houghton Mifflin. I worked with people my own age who were smart and funny and ambitious. I did well there, and when I came home flushed with my minor successes, the loft seemed dark and dreary. Maud was depressed and, increasingly, she depressed me.

I was offered the impossible after six months, a promotion to assistant editor, contingent on moving to the Boston offices for a year. It was a great opportunity, and I also suspected it was the only way I'd ever find to break free of my debilitating love for Maud.

My first month in Boston, I stayed in Cambridge, house-sitting for my new boss's neighbor while I hunted for an apartment. I still remember the smell of my landlord's perfume. It was cloyingly funereal and it seemed to have permeated everything, including my sinuses. Even when I left the house, that scent suffocated me. It also seemed like the snow never stopped falling that winter. All I did was work and come home, work and come home. I missed New York and I didn't like Boston. At night, I'd call Maud, who'd gone back to Nantucket. Each night our conversation would start off well and spiral into recriminations and despair. One night I couldn't stand it anymore and started to quietly cry.

"What's the matter?" Maud asked, suddenly alarmed.

"You know, I just can't take it anymore. I worry so much about you and about how depressed you are, and I just don't know what to do to help you. It makes me so sad."

There was a pause from her end, and then she said, with some exasperation, "Are you kidding? Come on. You know me. This is how I am. I'm always depressed."

And it was as though someone had lifted a rock off me. She was right. But I had no idea that *she* knew that—and once I did, it made all the difference.

Twenty years later, hardly a day goes by that I don't think of her and our jokes and our movie nights. Part of me still feels guilty, like I abandoned her after she summoned the bravery to face New York.

But I look back on this relationship with the knowledge that having saved myself, I will have the luxury of adoring her forever.

DAYDREAMING AT THE HELM didn't distract me from watching the instruments, checking the horizon and keeping an eye on the weather. Around 1:00 p.m., the sun faded and the direction of the wind changed. By 2:00 p.m. waves started to roll us gently from side to side and the autopilot required some attention to keep us on course but comfortable. Puffy cumulus clouds obscured most of the sky's bright blue by 3:00 p.m., and around 4:00 p.m. they had become more threatening cumulonimbus clouds. Another storm was on the way. This was a fairly predictable weather pattern each day, with a burst of rain and light wind moving through in the late afternoon and then dissipating. But today the skies got darker and darker, the wind got stronger and the rain didn't come.

We were in for some more serious weather. The cloud

bank chasing our stern rolled ominously toward us as John took the helm and I went below to secure loose items: the television, some glasses, books. Tightening the portholes, I chuckled when it occurred to me that I was actually "battening down the hatches"—it was the first time I understood the urgency in a phrase I had used so thoughtlessly a million times before.

When I was satisfied that everything was as secure as possible, I went back to the helm with our foul-weather gear. The dogs had already taken up their anxious positions. I think both John and I felt a little giddy. Adrenaline flooded my system as I contemplated the unavoidable emergency that was blocking our way. It was as though we were preparing for a siege, since there wasn't much we could do but brace ourselves and forge ahead. Maybe that's why I didn't feel scared. Fear comes with the knowledge that you can change your situation—get help, escape, overpower your attacker. We had no such options. We were about to practice the nautical equivalent of passive resistance. Our goal was not to overcome but to endure. I felt a sense of exhilarated fatalism. Hang on, here we go.

The sea was rough and the winds were blowing harder and harder, but the *Bossanova* was a champ. She held her own, bounding through the rough waves like an amphibious tank. Though it was quickly clear that the boat could handle this weather, John and I had never seen lightning like this before. Of course, the open horizon and dark sky provided a particularly ominous palette, and I couldn't shake childhood memories of grown-ups rushing us from the pool or pond when lightning started to flash. The dogs cowered in the corner: Samba was shaking from head to toe and Heck's beard was damp, a sure sign he was nauseous. John and I jumped every time a flash illuminated how very small we were and how very big the sea was, all around us.

For an hour, we tossed up and down on the gray-green waves, deafened by the grumbling thunder and sharp cracks of terrifying lightning, which were dangerously in synch. Rationally, I felt that we would be okay, that we wouldn't be hit and that even if we were, we'd survive. But this lightning was huge and getting closer and closer.

There's something particularly rattling about being in a small steel boat in a very big body of water with no other targets around. I wasn't sure that rubber-soled deck shoes were going to help if nature decided to fling some pyrotechnics our way. What I didn't know was that a steel boat was the safest place to be. Since steel conducts electricity, a lightning bolt would run through the hull to the ground (the sea) with little resistance. A wooden or fiberglass boat was much more dangerous—if lightning struck, the resistance of fiberglass or wood could potentially cause the bolt to blow a hole in the hull. In any event, a boat's electronics could be damaged by a strike. When lightning appeared on the horizon, a good emergency measure was to place a handheld VHF radio in your oven or microwave, which provided a natural faraday cage. If you were struck, your communications channel would still be working, making it possible to call out a Mayday as you sank.

Gradually, the waves calmed a little and the wind faded to a stiff breeze, even though the sea remained rough. The worst of the front just turned and wandered off, like an exhausted bully with attention deficit disorder. The only lingering threat was the lightning, and we watched a dazzling exhibit of it as we pulled into Obassaw Sound, Georgia, and dropped anchor. The chart indicated that this was about the best we were going to do in terms of shelter. Charleston would have to wait one more day.

That night, the wind whipped the waves against the steel hull, and the master stateroom, where the dogs and I slept, rang like a kettledrum. I tossed and turned and worried vaguely about the alarm not going off if we dragged our anchor. Every now and then I'd go up to the pilothouse and check our position against lights on the shore to be sure we hadn't moved. I had never used this GPS function before and I didn't trust it.

It was a relief when the sun finally came up, though I had barely slept. The waves had died down toward sunrise and the water was still and bright. We hauled up the anchor and headed back out, aiming for Charleston and a little R&R with the son of a Chapman classmate who ran a marina there.

I suppose many people would find life aboard a boat boring. I think John, who liked his creature comforts and was more outgoing than I am, sometimes felt a little caged and anxious. But my normally restless mind was quieted aboard the *Bossanova*. The little tasks of running the boat— checking the water flow in the saltwater strainers, keeping an eye on the oil pressure, scanning the horizon with binoculars, keeping track of our position on the chart and listening for Coast Guard updates on the VHF—were enough for me. I was vaguely aware that I spent much of each day with a goofy grin on my face. (I remember Captain Bob had once laughed at my perpetual smile and said that this was why he taught— to see the joy that every now and then one made-for-the-sea student would show.) From time to time the thought would pass through my head that I had no right to be running a ship like this, but the realization that I was actually doing it made me feel accomplished and proud. So I would rarely read while underway, preferring to stay focused on my boat, my command of our journey.

Sometimes, if John was on watch, I'd take a chair out on the bow and bask in the sun or watch the water for wildlife and other boats in the distance. But I never ventured far from the helm or escaped the present.

Today was another peaceful day with good weather and no mechanical problems. We reached the approach to Charleston around 3:00 p.m. John struggled against the usual afternoon waves that kicked up, getting progressively worse and becoming a strong beam sea as we got closer. We were eager to get ashore, and had plans to join our friend's son for dinner, but I decided we had to adjust course to min- imize our roll, even though that would take us out of our way. That helped postpone our rocking, but when we finally entered the channel, we had run out of avoidance tech- niques. I took the helm. We were really rolling from side to side now.

Complicating our rough ride through the channel en- trance was a wicked current running against us. We should have been making about 8 knots on our way in, but we were running around 3.5 knots. Not only were we being tossed from side to side, but we were going to be tossed from side-to-side for twice as long as we would be normally.

Samba and Heck were now each trying to occupy the same cubic foot of space in the very corner of the settee. It would have been comical if they hadn't been so terrified. (And these are two tough little Jack Russells, the breed they send into holes on the South African savannah to flush out hyenas!) I explained in soothing tones that everything was fine, but they were definitely not buying it. *Why couldn't we be in a normal house right now, with a normal yard, where we could chase normal cats and disobey a normal mother?* I imagined them wondering. Poor little guys.

Down below I heard a massive crash and said a small prayer that my tiny flat-screen television had not been broken. I snuck a glance behind me and it looked like backstage at the Steel Wheels tour. The salon was littered with my possessions: obviously, I hadn't done a good enough job of stowing and securing. The television was lying on its side, a good eight feet from where it should have been. If it still worked, it would be a miracle.

I finally felt my temper fraying. I was tired and I was fed up with fighting nature. John had the sense to steer clear for a few minutes while my seething turned to simmering and then blew over.

As we passed through the jetties, the waves flattened out a little, but we continued to combat the strong current. Charleston Harbor is one of the busier American ports, and we stayed alert as several large cargo ships passed us by, both coming in and going out. I braced us for wake and angled the bow into it as these vessels passed. We were fine.

Then, coming up behind us on our port side was a massive container ship from the Evergreen line. The aptly named *Ever Racer* appeared to be going much faster than the other ships we'd encountered. John got on the VHF and tried to raise their bridge but got no response. The *Ever Racer* just barreled by us. I was ready at the wheel, and John positioned himself to let me know what was coming. We had one large wave of wake. I steered us over it on an angle and we did fine. Then John shouted out that we had another coming our way. Again, we were ready, and it was big but manageable. John took another look and said, "Whew. We're clear, Mare," and then, as we both relaxed a little, we were suddenly up on the edge of something big that came from nowhere.

I was at the helm and doing my best but we were riding right on the crest and I could feel that I had absolutely no control. It all happened very, very fast, and there was no real time for panic. We were heeled over at a 35-degree angle to the water—John could probably have reached out and touched it!—and I thought, *This is it. We're going to capsize in Charleston Harbor.* I felt a stab of anxiety about the dogs, who were not wearing life vests, though it didn't occur to me to worry that we didn't have them on either. I had a quick mental image of the *Bossanova*, rolled so far over that she took on water through the pilothouse doors and swamped on her side. I imagined being towed to shore, the engine ruined by water. I even felt a flash of shame, though I knew it wouldn't be our fault. All of this ran through my head in a split second and I knew only luck might save us.

We did not capsize. Somehow, we managed to race down the face of that wake at a terrible angle and then right ourselves. The *Ever Racer* sped ahead without any indication that they had almost capsized us.

I was furious, absolutely fuming, and I have to admit, I couldn't let it go. I felt that we'd narrowly avoided the worst possible fate, through no fault of our own, and I wanted that boat to pay. I thought seriously about calling the harbor master and reporting the vessel, but in the back of my mind I feared we'd wind up in the nautical version of one of those trucker movies. You know, where the psychotic guy in the eighteen-wheeler is terrorizing a couple on vacation. I could just imagine us being chased up the coast by some salty freak-show who'd gone 'round the bend from all those years on the high seas. I decided not to risk it but I remained livid. It was hard not to be when we'd come so close to disaster.

Weeks later, when my anger had abated, but not my desire for revenge, I looked up some online facts about *Ever Racer*. She was part of the Evergreen Marine Corporation, a fleet of 150 container ships with a combined capacity of four hundred thousand 20-foot containers. The company's biggest claim to fame was a record fine of $25 million from the U.S. government for the deliberate discharge of oil waste into the Tacoma River. The bully who almost swamped us had a gross weight of 53,358 tons, and her speed was 23 knots. That may not seem very fast but the no-wake zone in a harbor is generally 5 miles per hour and a ship like this can take up to 4 miles to come to a full stop. So, you can see why any extra speed in a slow zone could be incredibly dangerous. I toyed with starting a campaign to have all boats wear big stickers on their sterns, like tractor trailers, with an 800 number for filing complaints. But out on the sea, might was more or less right.

Despite the near capsizing, we chugged safely into the Charleston marina and tied up at twilight. Unfortunately, we were accidentally put in a slip that needed an adaptor to deliver 30-amp power. One dockhand waited with us and I walked the dogs while another dockhand searched for an adaptor...and searched...and searched. They couldn't locate a single adaptor and were profusely apologetic for having put us in a slip with 50-amp power. They did have an open 30-amp spot on the other side of the marina, if we wanted to move.

It's difficult to explain. We really did need to recharge our batteries overnight. But I simply couldn't bring myself to fire up the engine, cast off the lines, move the boat into a new slip and tie up again. I just did not want to be on that boat for another minute that night.

The dockhands offered us some complimentary ice to help keep our perishables cold until morning. John and I said, "Sure, thanks," and sat in a silent, exhausted heap on the dock in the dark. This time, our close call had not left us feeling elated or triumphant. We were worn out. Depleted. Sick of the struggle to get from one place to another safely. After a while I said, "John, I can wait for the ice. Why don't you go grab us a table for dinner and I'll be right up."

I didn't have to ask twice. John got up and said, "Mare, I'll take you up on that. I really need to do some power-drinking but I'll save you a seat."

When I finally made it up to the marina hotel, it was after ten o'clock. I had dreamed of an uneventful arrival in the late afternoon, a great berth and a VIP greeting from a couple of dockhands waiting to tie us up and escort us to the marina manager, who had invited us to a barbecue with his family. The slip was not great, the manager was gone and the sunset barbecue was long over. Of course, the kitchen had closed, too, so there would be no dinner. But John had already made fast friends with the young bartenders, particularly the female ones, and was buying everybody shots.

When I parked myself next to John in a warm, safe place where one television above the bar was broadcasting headline news and the other showed a baseball game, I was happy. Not jubilant—just delightfully relieved. John bought me a shot; no Domaine Ott had ever tasted as good. Three days of narrow escapes had finally taken their toll, and John and I were both suffering from cumulative fatigue.

The bartenders arranged and rearranged a large squadron of single-serving airplane-size bottles on the counter behind them, just beneath a big plate glass window with a view of the harbor. Apparently, in South Carolina, this was the only bottle size that bars were allowed to carry. I could imagine that some good-old-boy who had a lock on the importation of minis also had a lock on a couple of key legislators. It was the silliest thing I'd ever heard of. And yet I ordered another one.

John was regaling our new friends (in an increasingly loud voice) with the story of our near-disaster earlier that evening.

"Yeah," said two of the barkeeps, almost in unison, "we saw a container ship coming through here today that we noticed was going *way* too fast." We didn't need independent confirmation, since we had almost rolled out there, but it was nice to know that these people, who watched the traffic day in and day out, had observed our reckless nemesis.

I decided soon after to call it a night and go back to the boat and the dogs. I slept like a log, and never even heard John when he returned. In the morning, as I made coffee and John emerged bleary-eyed from his stateroom, I suggested that we take a day off. John looked like he was going to weep with gratitude.

"That would be excellent," he said. "You read my mind. I just think that after the last few days, we could use a little R&R."

I tallied our hours underway and saw that we had run for seven hours on Wednesday, twelve on Thursday, seventeen on Friday, twelve on Saturday and twelve and a half on Sunday. No wonder we were fried.

After a long walk around the marina complex with the dogs, and lunch at the bar with John, I did some laundry, enjoyed a piping hot marina shower and sat by the pool in overcast weather to do some reading.

At one point, I looked up to see the *Ever Racer* heading back toward the Atlantic. Still infuriated enough not to care that I looked like a raving lunatic, I ran to the end of the pier, jumped up and down to attract the bridge's attention and then raised my middle finger defiantly. There. That would teach them to mess with me.

That night, I rejoined John at the bar. He hadn't left it

since they'd opened at 11:30 a.m. and he was in a great mood. He'd been watching baseball, drinking beer, and flirting with the pretty bartender from the night before. I feared that a certain amount of denial was at work, and that when we had to push off the next morning, John would cling to a piling and wail as we pulled away from the dock.

But he was stoic. He looked a little grim, and perhaps hungover, as he manned the lines and I backed us out at 0800 hours. We had conferred on the best way for me to maneuver us out of a rather tight spot.

The *Bossanova* has a left-turning propeller, which means the boat pulls to starboard when you are in reverse. I had this in mind as I planned our exit, of course. But the "prop-walk," as it is called, was augmented by the strong current, and suddenly our stern was swinging powerfully toward the dock. I had to turn the wheel hard to port and really hit the fuel to correct us in time. Some idiot on the pier shouted hysterically—as though that would help! But we powered away in time to avoid a collision with the concrete fingers. Still, it was extremely embarrassing and a tense way to start the day. Once we were back in the river, the strong current that ran against us on the way in was running with us as we left Charleston, so we were exiting the channel into the vast Atlantic in less than half the time it had taken us to fight our way in.

Neither of us felt sad to see Charleston slipping into the distance behind us. It was the scene of our closest call so far and our day off there had been more of an emergency Band-Aid than a fun vacation. It was good to be back aboard.

CHAPTER SIX

O, God, thy sea is so great, and my boat is so small.

—ANONYMOUS

Murrells Inlet, South Carolina. If you're ever in the neighborhood, stop by. Tucked back behind a strip of shoreline north of Pawley's Island and south of Myrtle Beach, it was one of the friendliest places we visited.

We had headed in a little early, after a blessedly uneventful day underway. The inlet was easy to run and opened onto a serene and gorgeous landscape—the late-afternoon sun setting lush patches of wetland ablaze with light. It seemed like we had entered another world, a secret spot that was amber and languid. After Charleston, which had been both difficult to get in and out of and very expensive, we were ready for a harbor that was distinctly not commercial. We had found it.

As we pulled into the little marina we'd found for the night, a small crowd of people came out to watch us tie up. Most of them had beers in their hands, but they were young and old, tattooed and preppy. A little boy hung around his father's knees and peeked shyly out at us. As I went below to make sure the battery charger was on and everything looked good in the engine room, I heard John talking to the folks on the dock. He had secured the lines and then immediately dipped into the cooler of Bud Light he kept on deck. Now he leaned against the hull with one hand and punctuated his conversation with the beer in his other. John knew the answer to every question about the *Bossanova* as well as I did, and it made me laugh to hear him—mostly because I could have sworn I heard a swell of proprietary pride behind his expansive responses. Listening to him now, you'd never guess that he was a fan of cigarette boats and Carolina-style sportfishers. Every time some slick *Miami Vice*–style boat whizzed past us, John would grow positively misty-eyed as he watched it fade quickly from the horizon. But I suspected he was starting to fall for the slow and shippy *Bossanova*. Maybe it was the sturdy way she'd carried us through some rough spots where a fast, planing race boat would have bounced around like a toy. Or maybe it was just another case of Stockholm syndrome.

It's hard to believe two people could spend as much time with each other as John and I did without getting a lot more intimate. Though we spent long hours in the pilothouse together, our unspoken agreement to avoid topics of dissension eliminated most conversational avenues. We whiled away the hours talking about Chapman pals and incidents, telling each other funny stories about our friends, planning our voyage strategy and listening to any kind of music we could agree on, which usually fell in that middle ground that neither of us loved.

The most personal things I ever found out about John were these: right after college, he got married and shortly thereafter divorced; and his previous career was so stressful that before he'd turned 30, he had high blood pressure and terrible insomnia. Both of these revelations surprised me. The John I knew now was a happy bachelor and as fun-loving as they came—a throwback to the Rat Pack. There was something innately good about John, something irrepressibly big-hearted that balanced out his many less-enlightened qualities.

But ours remained a friendship built entirely on our sea-faring adventure, our Chapman bond and mutual but affectionate scorn for our highly divergent world views. It occurred to me that maybe we recognized and liked in each other the will to walk away from other people's idea of success. We both knew how to relax and enjoy life, too. Whatever it was that bound us, it was working. We were getting along great, despite all our differences, long hours and constant companionship. On later, shorter voyages with people I knew much better, I found my nerves somewhat frayed by familiarity in such close quarters. John and I were a working crew, and we got along like two people who had a job to do but wanted to enjoy it as much as possible. It was oddly ideal.

While John continued to field questions from new fans of the *Bossanova*, I snuck off with the pups for a walk around the neighborhood, which was about one city block wide, with a main street running down the middle, ocean on one side and wetlands on the other. I sat for a while and looked out at the Atlantic—it seemed so different from this perspective. Here was a beach like the ones we visited on family vacations with our Hi-C and sandy sandwiches, where the seaside smelled like coconut suntan lotion, and the sound of the surf as you dozed off battled with the regular *ponks!* and occasional laughter of a couple playing Kadima. Here, the ocean was content to mute its destructive force by sending its quiet tide as an insidious invader of the sandcastle you'd worked on for hours. It was lovely, but it bore no resemblance to *out there*. A good reminder that it's all about perspective.

After I put the boys back on the boat, I joined John in the bar above the little store at the marina. It was a small, quiet, nondescript joint, with a magnificent view of the sun setting over the wetlands. Country music played on the jukebox. To my left was a guy of about 22, with spiky hair and multiple piercings. The bartender was a young lady with a midriff-baring shirt. The fellow at the end of the bar, to John's right, was an older man in a striped golf shirt. John had a couple of beers and I had a couple of vodkas, and we chatted with our barmates, who were, like people we met almost everywhere we went, intensely curious about the *Bossanova* and our journey.

When the sun had set, we thought of food and bed. I went home to the boat and the boys to see what I could throw together, and John decided to try the seafood place down the way. In the morning, we picked up a few staples at the little store and paid $40 for our dockage. It had been a wee spot of lazy heaven—easy to get to, laid-back and welcoming—a great change from our recent urban challenges.

Later that afternoon, the skies darkened just as we reached Cape Fear, and we decided to put in to Bald Head Island for the night. At first glance, we were happy with our choice. The island was home to a resort community of tastefully fancy houses—mostly large, shingled variations on Victorian. There were two restaurants—one formal, one informal—and a place to do laundry. There were no cars allowed on the island—everybody got around by golf cart. It seemed quaintly luxurious.

When we got up the next morning to another dark sky and a lousy forecast from the National Oceanic and Atmospheric Administration, otherwise known as NOAA, we decided to stay put and get some chores done.

Since we were low on supplies, John and I rented a golf cart to go to the island's supermarket. It was chock full of flavored coffees, $10 sandwiches with sundried tomatoes on ciabatta, embroidered baseball caps and organic angus steaks, but it was notably lacking in the small basics which couldn't sustain a 200 percent markup. The nearby hardware store sold expensive teak lawn furniture and huge stainless steel grills, but not the batteries we really needed. Just one extra day was wearing away Bald Head Island's scenic patina and revealing a place that seemed more like a stage set than a real town. Suddenly, it seemed that every other golf cart conveyed a blond family in madras shorts and a golden retriever (who was invariably named Max). I found myself pining for the pierced kid from Murrell's Inlet. That afternoon, when the skies cleared and the rain stopped, I felt restless. I tried to focus on being productive, but the day had a wasted quality. I didn't want to be hanging around an expensive marina, twiddling my thumbs. I was ready to go back to sea.

The following morning, the weather was lousy again. We listened to a marine forecast that promised intermittent storms all day, but we decided we just didn't want to forfeit more time. We agreed we would go out but stay close to land and keep a constant watch on the weather.

A few miles offshore, we realized we'd made a mistake. The sea was rough, the skies were dark and we were going to be running in the vicinity of the infamous Frying Pan Shoals, a long line of submerged but treacherously shifting sand and rock that extends 20 miles offshore and sweeps up the coastline. Discretion is the better part of valor, after all, so we hightailed it back toward land. Our compromise was to run in the relative safety of the Intracoastal Waterway for a day or two.

Not long after we turned back, we heard the Coast Guard on the VHF. A pleasure boat named *Spring Fever* with a family aboard was taking on water off the Frying Pan Shoals. The Coast Guard asked repeatedly for latitude and longitude but the boat's GPS was down. We could imagine how scared they were, 15 or more miles offshore in lousy weather, in need of immediate assistance but with no real idea where they were.

In a situation like this, the Coast Guard has a very specific drill. They run through a series of questions for the captain in distress: *What is your exact location? Could you please describe your vessel? How many people are on board? Is anyone injured? Is everyone wearing a PFD* (personal flotation device)*? Are you or your vessel in immediate danger? How much water have you taken on?* In this instance, as in others, we could sense the frustration of the boat owner, not from what he said (because we could only hear the Coast Guard's side of this conversation) but because we could hear the Coastie trying to calm him. *Sir, I understand your GPS is not working. But do you have an approximate idea of where you are? Could you check the GPS again? Is it still not working?*

Okay, first, let's admit our sinking boater was an idiot if he knowingly set out with a broken GPS and without taking bearings to mark his position on a paper chart. But the Coast Guard dispatcher's persistence was maddeningly robotic. If he didn't get an answer to any one of these questions, he just stayed on it, asking repeatedly. I later recognized that there was probably a method to this madness. The dispatcher is maintaining constant communication and asking for specific information from someone who might otherwise be hysterical or trying to do too many things at once. On the other hand, it's probably infuriating to describe the exact color of your vessel as you watch her hull sink beneath the water...

John and I were amazed by the distress calls we heard over the VHF. It was both compelling and upsetting because these were very real emergencies at sea, and never very far from our location. We heard something dramatic almost every single day we were underway. There was a 90-foot fishing vessel that had taken on four feet of water in the engine room. There was a man overboard. There was a pleasure boat swamped at an inlet. There were several Maydays that the Coast Guard sought more information on after losing all contact. There was a fire. There was a captain who had a heart attack at sea. We even heard the Coast Guard alert mariners to a plane that had crashed into a bay. John and I took bets on what could possibly be next: an outbreak of shipboard leprosy, a spontaneous combustion, a boat being used as a crack house?

It was frustrating that we often had no idea how these emergencies were resolved. Imagine the suspense: it's a real life-and-death situation, but you don't get to know the ending.

VHF action aside, the *Bossanova* and her crew had a gray and dreary day on the ICW. We ran from 0815 to 1800 hours, taking turns at the helm, as the day slipped slowly away, then anchored off the main channel near Mile 246.

IT'S ODD THAT THE longest love relationship of my life is also the one I think of the most dispassionately. Or maybe that's not odd at all. Laura and I had probably wrung every ounce of emotion from our partnership long before we came to an end.

I was 28 when we were set up on a blind date. We met at a cozy bistro in the West Village where we drank too much red wine and ate pâté. She was three years younger than I, pretty, a prep school and Ivy League graduate, a champion athlete, well-read, relatively up on current events. What wasn't to love? We hit it off and six months later, she gave up her charmless studio in Brooklyn and I gave up my charming studio on Perry Street and we moved in together.

Laura was a struggling singer-songwriter and a highly paid tutor on the side. She really didn't have much money, but like many performers, she pampered herself: healthy foods, long sessions at the gym each day, summers in the country. On the one hand, she professed to adore me. On the other, she would often claim she'd kill herself if she didn't make it as an artist. It's not a statement that falls on the beloved's ears ringing of adoration. I tried to make her see that her life was happening right now—not when she got a record contract—and that it was really pretty damn good.

We had close friends and family. We had a dog we were both besotted with. We had weekends in the countryside where we cooked great meals, watched movies, went hiking. And, oh yeah, we loved each other. I thought that should have some small value, at least.

I did everything I could, of course, to help Laura professionally: rallied my friends to attend every show, helped address invitations, used whatever contacts I had to get people of influence to hear her. She wasn't the next Joni Mitchell, but Laura had as much talent and dedication as many people who make it—it just didn't happen for her. And she had a stardom clock ticking away just the way some women have a baby clock. Laura believed that if she didn't make it by the time she was 40, she wasn't going to make it, and 40 was no longer terribly far away.

I knew I could never replace the success she wanted, so ours was a strangely unintimate love—sometimes companionable, sometimes stormy, but it seemed as though all of our emotional exchanges, whatever their flavor, happened across a chasm of space between us. To be fair, I didn't react well to Laura's increasingly hysterical demands on my attention. I had no idea how to make her feel whole and I grew to resent the fact that my best efforts were obviously not good enough.

I realized at some point that though she loved many things about me—my sense of humor, my career, my looks—Laura really didn't love *me*. She loved a hologram of me, and that was why I felt so lonely, so far from her.

We broke up after four years. Three months later, we got back together. Things were different, though not necessarily better, and two years later, we broke up for good. Six years of my life, and until last year, when I saw her, I felt only friendliness, kindness, a topical interest in how she was doing—not that I didn't go out of my way to help her, subletting her my cheap apartment when I went to Pennsylvania, giving her comments on her proposal for a book about not making it as a singer-songwriter and introducing her to agents.

I could say more about the lousy things she's done since then, but I will take the high road. I still don't understand how I could have spent so much time with someone who clearly cared only for herself. Never mind. I don't even want to know. The main thing is that although I will no doubt make plenty of other mistakes in my life, falling for a narcissist won't be one of them. Thanks to Laura, I've got antibodies galore.

THE NEXT DAY THE weather was better, and we continued through North Carolina, up the ICW past Morehead City and Beaufort, which both teemed with small watercraft. My good friends Frank and Barbara Sain had offered me a free mooring in Beaufort and I'd heard it was a wonderful town to explore. But we passed by far too early in the day to justify stopping, especially after two days off. We'd now been under-way for twelve days and we weren't even halfway to Sag Harbor. North Carolina was starting to seem endless.

We took Adams Creek, the cut that runs between Beaufort and the Neuse River, and enjoyed the smooth ride of the big channel. We saw two huge tugs with barges, but almost nobody else. Toward the end of the afternoon, we opted for an anchorage that one of the smaller guidebooks mentioned. It was off Adams Creek, at Green Marker 9. The channel mark-ers in that area had been moved around, but a range marker was clearly indicated on the chart and still exactly where it should be. So we knew we were in roughly the right place.

The anchorage was tricky to get into. No one else was there, and it was littered with fishermen's floats, so we pro-ceeded carefully, checking the chart and choosing a bearing off the lower range marker that showed plenty of depth. Once in, we tried repeatedly to anchor but the bottom was nothing but a powdery silt that wouldn't hold our hook. Defeated, we decided to get back into the ICW and head for Oriental, since we still had plenty of time before dark.

I was at the helm. As I turned us around, I glanced at the crab pots again, looked for the range marker and angled back the same way we had come in. Or so I thought. We were making our way out slowly but confidently when I suddenly thought, *Uh-oh. Did I choose my bearing off the low range marker or the high range marker?* Unfortunately, the nearly simultane-ous shudder beneath our boat indicated that I had aimed for the high range marker. We were aground. Really aground. And this time, no amount of powerful maneuvering could free us. For fifteen minutes I revved the engine and turned the rudder—hoping that what had worked in New Smyrna would work again, but we were well and truly stuck. Overconfidence had made me casual in a situation where I should have known better.

At first, I kicked myself. Part of it was ruining a good record. Of course, we could sit and wait for the tide to come back and float us free but the days off had put us behind schedule and we didn't want to lose more time or end up trying to find a safe anchorage in the middle of the night. I called Towboat U.S. on the VHF, and John and I decided to relax while we waited. I was only mildly annoyed with myself. Nobody's perfect and I had been a little careless. It was a very valuable lesson learned and luckily no harm had been done.

As we waited, a rowboat with an old man and a teenage boy approached us and circled curiously. "Hey, you run aground?" the younger of the two occupants called out.

"Yup, looks that way," replied John good-naturedly. "Our own fault."

"Yeah, well, happens all the time in here. You need a tow?"

"We do but we've already called Towboat U.S. Thanks a lot, though" said John.

"You sure?" The young guy in the rowboat tried again. "We could pull you off faster in *Shrimpboat U.S.*," he said, flashing a gap-toothed grin, "probably cheaper too. $50. How 'bout that? We could do it right now, no problem. You'd be on your way."

John and I had seen a fleet of big shrimp boats, lined up in the adjoining cove. But I was glad I'd already called Towboat U.S. For once, I hadn't let my optimism get the better of me—I had chosen the unlimited tow option on my insurance, so being pulled off wouldn't cost me an extra dime. But more meaningful than the money, Chapman had drilled into us the danger of offering or accepting a tow from another vessel. In the wrong hands, it could turn into a disaster. A poorly chosen line will sometimes break under pressure and cause a serious injury when it snaps. Or a cleat will tear loose, damaging the boat and hurling a metal object with great speed and force at the nearest bystander. If you're a good Samaritan who tows a boat in trouble, you could nonetheless wind up with a lawsuit for any injuries to the other guy's boat or passengers, whether you caused them or not. The same is true if you accept a tow from a friendly bypasser, of course.

And that's even before you got into the issues of salvage towing. Under maritime law, there are incidents in which a troubled vessel at sea becomes the property of its rescuer. Likewise, a ship that has been abandoned—even for reasons of life and death—can become the property of whoever tows it to shore. Some boaters vacationing in less developed parts of the world have gone ashore for lunch only to return and find their unattended boat has been cut from its anchor line and set adrift so it can be "salvaged." It's always better to use a professional towing service if you aren't in dire trouble.

We waved as our locals rowed off in disappointment. And I wondered if they had perhaps moved the markers around to bring a little extra business into their part of Adam's Creek. Was that banjo music I heard? Before my imagination could get too carried away by what a night trapped aboard in this anchorage might be like, we heard Towboat U.S. calling to say they were a mile away.

A few minutes later a red towboat entered the cove and circled us. The mustachioed driver introduced himself as John Deaton and explained how he intended to get us off the bottom. At first we were incredulous. I had given the dispatcher the *Bossanova*'s measurements, emphasizing that she was steel and weighed 30 tons. But the boat that circled us now looked tiny—more like a ski boat than a tow boat. It was about 20 feet long, fiberglass, with a covered center console and big twin engines. A large stainless steel pole was set through the aft deck, obviously for tying to a tow.

I expressed some doubt, and Deaton grinned. "Don't you worry," he said. "These engines are very powerful, and before I tow you, I'm going to use them to float you loose. Watch this."

He tied us to his tow pole and backed right up alongside us. With expert turns of his boat and big engines, Deaton started blowing sand out from underneath the *Bossanova*. It was not, I suppose, unlike how I'd gotten us off the shoal in New Smyrna, but the engines were much more powerful and directed. Deaton worked for about five minutes, and then we felt ourselves floating free again. He pulled us forward about 100 yards, then had us cast off the tow line.

"Ok," he said. "You follow me out of here. Where you folks going anyway?"

We said we intended to stop in Oriental for the night.

"Good idea," he said. "That's not far, about 5 miles ahead.

There's a big fireworks display tonight for the Fourth of July. We do ours a night early. It's going to be awfully crowded, but you can probably find a spot to anchor and watch. I'll make sure you're okay getting out of here, then I've got to run on back to town. I'll have a look and let you know if I see any space for you. We can handle the paperwork for the tow back in town, okay?"

It was fine by us. As we approached Oriental about 45 minutes later, John Deaton came back on the VHF. *"Bossanova, Bossanova this is Towboat U.S. Come back."*

"Towboat U.S., Towboat U.S. This is Bossanova."

"Yeah, Bossanova. Switch and answer Channel 9."

"This is Bossanova, switching to Channel 9."

Once we were on a less-trafficked frequency, John Deaton came back and said, "Yeah, *Bossanova.* Here's what's happening. There doesn't look to be any space at all in the harbor. But me and my brother have a marina, and we have a spot for you for tonight if you want to tie up there. There's no toilet or shower, though. Just a place to dock."

We had a shower and toilets with holding tanks aboard, so there was no problem. I hoped it wouldn't cost us a fortune, but I didn't feel like we had a lot of options this late in the day.

Deaton gave us directions. We had to look for a narrow channel off to port, shortly after we entered the harbor. Follow the markers carefully, staying to the starboard side because it was a little deeper there and we were *just* going to make it, with our 4-foot 9-inch draft. At the end of the channel was a kind of tricky turn to starboard and then almost immediately after that another hard to port. We should take the third slip on the right.

His directions were good and the channel was exactly as he described. At the end, where the short tricky turns began, was a clubhouse of some sort with a big deck. There were plenty of observers celebrating Saturday afternoon on the Fourth of July weekend as I brought my little ship perfectly through the turns and smoothly into the dock. Even I was a little impressed with myself as people shouted out *Beautiful job! Nice work!* It was one of several moments on the trip when I recognized the pride I felt at becoming competent. (Of course, these people didn't know I had run us aground about two hours earlier.)

Over the years, I'd seen books I edited stay on the *New York Times* bestseller list, get great reviews, sell as many as 10 million copies. This satisfaction I felt from bringing the boat in flawlessly was infinitely more real and thrilling. It seemed perverse, but as Emily Dickinson first said, and Woody Allen said much more infamously, the heart wants what it wants.

We tied up in our slip across from the clubhouse, and a whole new group of onlookers approached to congratulate me on my boat handling and to ooh and ahh over the *Bossanova*. We were essentially bow-to John Deaton's brother's backyard, and he was having a big party. Children were running around playing hide-and-seek, a group of guys stood around the barbecue, women were going in and out with covered dishes. Everyone was holding a beer. The scene was very Americana and oddly soothing, though far from my idea of family. It was too traditional: the men were relaxing in a kind of macho way, talking about NASCAR and swearing frequently. The woman all looked a little damp and harried as they cooked and supervised the kids. John was right at home though, drinking beer with the guys in no time flat. I sat with Deaton while he filled out the towing paperwork.

When he handed it to me, I had to stifle a gasp. That grounding and short tow had cost Boat-U.S. $750. Wow! The towing business was *all right*. I signed, handed the clipboard back, and thanked him again for finding us this spot. What did we owe him for the night?

"Oh, there's no charge," Deaton said. "Slip would just be sitting here empty, and it's not like we have any services to offer you. Just enjoy the fireworks."

That was a generous gesture and a nice surprise. I smiled when a gang of kids showed up and piled into the boat next to us. They were lugging a big picnic basket and shooed ahead by a pretty woman who turned out to be Deaton's wife. *Yeah, we thought we'd take dinner, anchor out and watch the fireworks from a less crowded place,* Deaton explained. That was the family I'd want to belong to.

John and I did indeed sit on the stern and enjoy the fireworks. They weren't much, actually, but we were as happy as could be seeing the colors explode against a sky full of stars while we held paper plates on our laps, piled with barbecued chicken and baked beans. This wasn't too bad at all.

The next morning, we decided to head for Ocracoke Island, part of North Carolina's famed Outer Banks and home of Edward Teach, the fiercest pirate of them all, better known as Blackbeard. Actually, Blackbeard's legend may be a good deal fiercer than he was, but that's the way he wanted it. With long dark hair, a bushy beard and an array of daggers and pistols swinging from his bandolier, Blackbeard projected an intentionally savage image. He was often seen with smoke curling from his ears—said to be the result of slow-burning matches he placed in his braided facial hair before a battle. Legend has it that on the last night of his life, Blackbeard learned that the British Navy was coming to the island to capture him. Unable to navigate the treacherous shoals in the dark, Blackbeard paced the night away, yelling, "Oh, cock crow! Oh, cock crow!" willing daylight to come and thus giving the island its name. Daylight did not arrive soon enough for Blackbeard. He was captured and beheaded by the British. Legend has it that his headless body circled the British Navy's boat many times before he died, though many hold that he did not, as they say, give up the ghost. There are reports that his spirit body is often seen swimming in Ocracoke Inlet, searching for his head.

Luckily, we didn't see it on our way in but we did witness the famously tricky shoals that spelled his doom. We had planned what seemed a safe route through, but as we approached we saw a large passenger ferry coming up behind us.

"Hey, John. I'm sure our route is fine, but why don't we take advantage of this guy's local knowledge and see if he'll let us follow him in?" I suggested.

John got on the VHF and offered to pull over and let the ferry pass if we could follow him in. No problem, said the ferry captain. And we did. The ferry took the exact same route we had planned, though it was faster than we were—who wasn't?—and we lost sight of it as we neared the harbor. I'd read something in one of the guides about overshooting the obvious entrance and then doubling back at a sharp angle to avoid some new shoaling. But John suggested I should just turn and make a straight approach at the entrance. I'm pretty sure that's the way the ferry went in, said John. It'll probably be fine, I lied to myself.

We touched ground for a second but I powered us off. I left a big billow of red behind and panicked, wondering if I had chopped a nurse shark or sting ray or some other large marine mammal. It was much later that I realized that the fancy bottom paint that's designed to shed barnacles and marine growth with ease sheds itself as easily. That had been a big billow of expensive bottom paint I felt terrible about killing. I had a good laugh, and not for the last time, at what I still didn't know about boating.

Once in Ocracoke, we found the municipal marina right next to the ferry dock. We sat in the busy harbor and tried repeatedly to raise a park ranger for an indication of where to pull up, but we got no response. Well, I figured, let's just go in and tie up anywhere, and we can move after that. Bobbing where we were unnerved me, particularly as I did not have a chart with enough detail to indicate water depth inside the harbor, though boats buzzed by all around us. Touching ground on the way in had made me a tad paranoid about taking anything for granted.

We entered the municipal docks and I began to turn toward a temporary spot alongside the long pier when a berthed trawler hailed us on the VHF.

"There's a spot over here that's free if you head straight back and then turn to port. You won't be allowed to stay alongside that pier," he offered.

"Thanks for the info," John called back. "We'll give it a try."

There wasn't a lot of room to turn once we were in the dock area and I had already started angling us toward the pier. Now I tried to turn us back in the other direction, but the wind was against us. The *Bossanova* was a great ship, but she had a big profile, a lot of windage, and she was tough to dock in a stiff breeze. After a moment, I realized I was going to have to make a very tight turn to port and go back to the original plan.

It went smoothly enough with me using the technique Captain Bob Swindell had drilled into me: big, short surges of power. But the turn I needed to make today was difficult with the wind pushing us forward, so I had to alternate between surging ahead and surging back to keep us pivoting in a small enough space. What I was trying to do was wind up with my stern toward the dock so the wind could push us in and alongside. Anyway, that was the idea.

I was controlled but tense as I executed this plan since there was absolutely no margin for error in such a tight space. Also, docking the *Bossanova* always created a bit of a spectacle. She was a unique little ship and her drystack exhaust gave her a throaty, chugging sound that I loved. Using a lot of power tended to amplify that, and it was hard not to attract attention. I loved it when people stared appreciatively at my boat, but *not* while I was docking. It made me much more nervous. What had happened in Oriental with the cheering on-lookers had been a lovely exception—the last thing I needed in this tight, windy situation was people watching me.

Adding to my stress level was a bright red fiberglass sail-boat, about 26 feet long, also tied alongside the pier and directly in front of my bow as I worked on the turn. Every time I surged forward I came unavoidably close to the little boat. I wasn't worried about what would happen if I hit a dock or piling with the *Bossanova*. She was a tank. But the last thing I wanted to do was crush a small sailboat while her two owners looked on, terrified, from her deck.

But luck was with me. I slowly powered us around, then the wind caught us and pushed us right up against the dock. Before we could even tie up, and well before a sense of relief overtook the adrenaline that still coursed through my veins, Heck and Samba leapt from the boat onto the dock. On more than one occasion, they had pulled this escape trick. I had much better visibility with the pilothouse doors open, but if I tied them up they'd become hysterically excited and yappy, which made concentrating completely impossible. So I left them loose, and though I kept netting across the gap in the railings, they sometimes leapt right from the side of the boat. It made me want to strangle them because the last thing I needed to think about was chasing those two down when I had a 30-ton boat to finish securing. But I tried to see it from their perspective: the sea was not their friend. They saw a chance and seized it. My advice: never make the mistake of getting dogs that are smarter than you are.

A couple of bystanders grabbed my furry fugitives and the

park ranger finally appeared to say we were welcome to stay where we were for a few hours, but then we'd have to move. The new spot we were assigned was pretty easy to get to, so we untied and reberthed immediately, then hit the town.

Ocracoke was hopping. The outermost of North Carolina's barrier islands, Ocracoke is only accessible by ferry or private plane, and its population swells from about 750 in the winter to more than 7,000 each summer. At one end of the island there's an adorable, historic village full of cafés and shops, but the main draw is the 16 miles of windswept Atlantic beaches, with plenty of room for both wild ponies and tourists. Despite a relaxed atmosphere that suggests time has passed it by, Ocracoke is under a lot of strain, and not just from the massive influx of visitors: the beaches are eroding at a rate of 10 feet per year and the small island would more or less disappear if hit by a large enough hurricane.

But today, Ocracoke's worries are invisible: a raucous Fourth of July parade with a half-dozen floats had just made its way up the main street. Kids chased the end of the parade, honking horns, waving American flags and cheering. As I walked the dogs along the dock, an older boater who was sitting on the flybridge with his wife called out that I had done a beautiful job of docking. I was appreciative but admitted my anxiety about such a tight turn in that wind. Well, he said, it didn't show.

When the dogs were back aboard, John and I decided to look for a fun place to celebrate. On the way up the sandy main street that runs along Silver Lake Harbor, we decided to stop at the beautiful British cemetery.

Ocracoke's history dates back to 1500, when the Ocracoke Inlet was first used as the lane to Pamlico Sound and the North Carolina coast. But it has an interesting modern history, too. In 1942, American ships were busy patrolling the eastern seaboard, but only one American ship was sent to protect the southeast coast. Eager to defend important supplies bound for England through U.S. shipping lanes, Churchill sent a flotilla of antisubmarine craft. The HMS *Bedfordshire* was torpedoed by a German U-boat on May 11, 1942, and all hands were lost. The bodies of four British sailors washed up on the island, and the people of Ocracoke declared the land where they are buried to be British soil. Each year in May there is a ceremony to honor these dead, attended by officers of the British Royal Navy and the U.S. Coast Guard. The famous Rupert Brooke quote marks the spot: "If I should die think only this of me: that there is some corner of a foreign field that is forever England."

The sound of Independence Day on the Outer Banks— firecrackers exploding, kids yelling, horns honking—sucked us back to the present once again. We returned to Harbor Road, with its cute cottages and restaurants, in pursuit of a cold adult beverage.

We decided to try the first waterside joint we stumbled across, where the Jolly Roger flapped in the soft breeze. The proprietor should have been keel-hauled for the bad crab cakes, but we had a great time relaxing with cold beers, then crossed the sandy street to a little upstairs bar that had been furnished like somebody's idea of a living room in Tiki paradise. It was kitschy and cool, but it must have drawn a late-night crowd because absolutely no one was there. We had a drink and chatted with the bartender, then tried a tropical-looking but dark bar that advertised a dozen shrimp for $3 at Happy Hour. As we sat with drinks and waited for our order, John braced himself against the bar with both arms and said, "Whoa. This is very weird, Mare. I can't stop the sensation that I'm rocking." I can testify that his sense of undulation was not at that time related to overindulgence. I'd heard of this happening to sailors before.

I left John after a couple of iffy shrimp and started back to the boat, noting on my walk that what the town lacked in haute cuisine, it more than made up for in old-fashioned charm. Since it was almost dark, I took the dogs up to the park, and we sat on the grass to watch the Ocracoke fireworks display that their historical society sponsors each year. They were absolutely gorgeous. There were about six surges toward the end, each one worthy of being the grand finale, but every time I thought it was finally over, there would be yet another spectacular display. I wondered how a little town like this could fund such an extravaganza—it was one of the best pyrotechnic displays I'd ever seen, and I'd seen some George Plimpton–narrated Grucci Brother beauties. Maybe it was just the setting, but I had to give it to Ocracoke Island: they know how to celebrate the Fourth of July better than anywhere I'd ever been.

The next morning, we got underway fairly early, but we couldn't get the computer to work once we'd left the dock. Loaded with digital charts and navigation software, it was our lifeline. And to make things worse, I soon discovered that this was the one portion of our trip for which I did not have a paper chart backup. It was not until after we'd arrived and were wandering through some T-shirt shops that I learned that the water off Ocracoke's coast is nicknamed the "Graveyard of the Atlantic." Over a thousand ships have sunk in the waters off North Carolina. In other words, there are better places to discover you don't have a nautical chart.

The computer had been working fine before we left the dock, but nothing I did now could get the display working again. There was no choice but to turn around and go back. I wasn't going to risk running these treacherous waters without a chart. Hopefully, I'd be able to find one back in town.

It was blisteringly hot outside and I noticed that John was looking a little green around the gills after his extended Fourth of July celebration. When we were once again tied to the dock, we walked up to the air-conditioned splendor of the nearby tourism office to use the soda machine and ask for some advice. The attendant looked doubtful but mentioned a few places that might carry charts. John went back to the boat, and I walked into town with my fingers crossed. I wistfully eyed a couple of $3 laminated placemats with maps of the island on them, but I was pretty sure that points marked "Here be treasure" and "Thar be sea monsters" were indications that these might not be accurate enough for navigational purposes. I wound up paying $45 for a paper chart that would be useless once we were 5 miles offshore, not to mention $30 cheaper anywhere else.

Back at the boat, and already drenched in perspiration at 10:00 A.M., I fired up the engine again and for the heck of it tried rebooting the computer one more time. It worked, of course. I decided I could have a temper tantrum about wasted time and money or I could just be grateful that I had my software and charts back and wouldn't have to replace the whole shebang. I loved it when I talked some sense into myself.

We were now ready to go and I asked John to prepare to cast off the bow line. I heard a groan.

"Cap, I'm sorry. I don't think I can do it. I just threw up over the side. I was fine when we went out this morning, but when I got back aboard, everything started rocking again." John turned and heaved over the railing again just to drive his point home.

I was disappointed, but there was no doubt in my mind we should stay. John was making this long trip for the sea time, without pay, and I knew he was getting more than he'd bargained for. The least I could do was wait for him to feel better. I was pretty sure that a massive hangover and the disgustingly hot day were adding to his rocking sensation, and I suggested he indulge himself in some creature comforts. I didn't have to ask twice. John promptly headed to a nearby hotel for air conditioning, ESPN and lots of Coca-Cola. The boys and I paid for another day at the dock (which was delightfully inexpensive!) and made the best of it with open portholes and multiple fans.

The next morning, John showed up in excellent spirits and we set our course for Oregon Inlet, the next good place for us to get to the outside again. We were following the almost universal advice to run inside Cape Hatteras, not outside, where the weather could be dangerously unpredictable.

Oregon Inlet was at the northernmost end of Cape Hatteras, though, and seemed like a safe place for us to exit back into the Atlantic.

Shortly after we were underway, we heard the Coast Guard broadcasting to all mariners to be alert for a man overboard. He'd gone over the side of a vessel called *Wild Man*, 37 miles southeast of the Oregon Inlet at approximately 0630 hours. All day long, we listened to the same information, repeated every half hour or so. Our hearts sank as the day wore on and the search continued. When we tied up at 1800 hours, the Coast Guard was still broadcasting the same information.

We motored in to Oregon Inlet Fishing Center as the sun was starting to sink. The entire marina was already full of big, beautiful sportfishers—a fleet of boats with tuna towers and fighting chairs, charter boats for those seeking big fish. They glistened in the tawny twilight, all lined up and tucked in for bed. We got the very last berth. When I took the dogs for a walk and glanced back at *Bossanova* in her slip, it was hard not to play "Which of these things is not like the other?"

I'm not a big fan of sportfishers, but I had to admit these boats were sexy. Sleek and shiny, with a dramatic curve where the hull meets the deck known as a Carolina flare, each of them cost well over $1 million. The fishing center was where the entire charter fleet docked, about fifty boats, I'd guess, and where they sold their extra catch at the end of the day. We watched as local people pulled off to the side of the road and ran across to the dock. A man was bent over a large fish with a flashing blade and a small throng waited to pick up dinner. When he finished cutting, the crowd had dispersed and the entire fish had vanished in under three minutes, except for a slick of blood that was soon hosed off the dock.

The next morning was beautiful, and John and I were dying to get back outside and make up for lost time. We'd spent the better part of a week in North Carolina and—no offense to the tarheels or their lovely state—we were sick of it. Some of the joy had gone out of the journey in the last week, but I realized later that, for me, the frustration was due to not being offshore or underway. There'd been too many days off, punctuated by the bad weather that forced us into the ICW. Today, we were rarin' to go and determined to leave North Carolina and our malaise in the *Bossanova*'s wake.

Yesterday's man overboard was still on our minds, and we monitored the VHF all day, hoping to hear some news. When we didn't, we decided that maybe it was a good sign—they'd stopped looking for him because he'd been found. But months later, as I read *Soundings* magazine on a Maine fall day, I saw a short item. The Coast Guard reported a man overboard off Oregon Inlet who was never recovered. I didn't really need to, but I checked the dates against my logbook. That was our guy. It still made me sad.

Our run toward Virginia was uneventful until midafternoon. Ten miles south of Virginia Beach, we watched a wall of dark purple clouds turning black. It stretched from the gunmetal gray of the ocean's surface endlessly into the sky. And it was coming for us.

We'd been monitoring Channel 16 as well as the marine weather channels and there had been absolutely no warning of a change in the weather. This looked much, much worse than the storm we'd encountered just past Jacksonville.

With exquisite comic timing, a young and nearly unintelligible Coast Guard voice called out from the VHF: John and I had already observed that these young Coasties tended to read their bulletins as fast as they could, usually in a heavy regional accent and with as little modulation or enunciation as possible. Kind of amazing, given the importance of the information they conveyed.

'Tention all stations, tention all stations. This is U.S. Custard, Ginia Beach Station. U.S. Custard, Ginia Beach Station. A seveah storm warning has been sued for the area between noath thitysixdegreesfittyzerominutesandzerothreahseconds and we-ehst sentyfivedegreesfittysenminutesandzeroayuhtseconds and will be mvving into this ayah in the next teun to fifteun minutes. There will be heavy rains and lightnin. Windsill be from thirty to fohty miles per hour with gustsupto fittyfive miles per hour. All mayiners are advised to seek safe harbor immedletly.

We had to laugh. Nervously, of course. Yeah, sure, the storm is coming in ten to fifteen minutes and we're supposed to find a safe harbor when we're 5 nautical miles offshore and our cruising speed is under 8 knots. We were clearly not going to outrun this storm, but maybe we could outsmart it.

John and I agreed that the first thing we should try to do is get out of the way. Since the storm was coming at us head-on, we turned and ran south and east, away from it and off its leading edge.

We both gazed with real awe at the massive thunderheads behind us, and I kicked myself for not buying a video camera.

Our marine weather teacher at Chapman had been an enthusiastic young guy, new to teaching, who loved his subject and delighted in bringing in great weather photos and satellite images to enliven the class. We had seen some amazing pictures of storms in action. But this looked as bad as any cloud he'd ever shown us, if not worse. Maybe it just had a lot more immediacy while stalking us at sea than it had when projected onto a slide screen in Johnson Hall, but this was very, very ugly. Huge, gunmetal gray and violet, it rolled toward us, doubling over on itself like some kind of diseased cell, spreading its contagion.

The rain started in slow, fat drops that turned to a downpour within seconds. It tattooed the surface of the sea, changing it from green to gray and drawing a blurry curtain across our visibility.

Our escape strategy was at least partially working, however. We had managed to skirt a good chunk of the early part of the storm, but it was blossoming now, billowing eastward across the dark seas and definitely gaining on us.

"Are you thinking what I'm thinking?" I asked John. "I have this crazy idea."

"Wait," said John. "Are you thinking that we should turn and run *into* the storm now? I was just wondering about that, too."

It really wasn't as foolhardy as it sounds. It was clear that as the storm spread out, we were not going to outrun it. The mouth of this monster—led by a solid black pillar of rain—was now nipping at our heels. And it was going to catch us soon. If it continued at its current pace, but we ran directly into it at our current pace, we'd effectively halve the duration of our suffering—and that seemed about all we could hope for.

We turned and headed *Bossanova* back toward Virginia Beach and into the belly of the beast. Samba, already in trauma mode, had gone below to hide in my stateroom, shaking uncontrollably. Heck was wedged into the corner of the pilothouse settee, curled in a tight ball and visibly nauseated.

The wind was fierce now and we were heeled over about 15 degrees and being pushed off-course by about 15 degrees, too. Waves were between 6 and 8 feet, but we were taking them head-on with no problems at all. Thank God we'd taught ourselves the good habit of entering our destination coordinates on the GPS, because it continued to feed us course correction data while our visibility deteriorated to almost nothing. The radar also continued to flash a reassuring outline of the coast, but nothing else in the vicinity. We could see nothing at all—we were running blind except for our electronics.

The worst of the storm must have lasted about two hours, though time seemed to stand still while we were in the thick of it. I suppose we were completely focused on the here and now, utterly absorbed in surviving the moment. When you think about it, there are very few occasions in life when the mind isn't free to roam forward or backward at least a little bit, while the present is briefly put on autopilot. Pitching around in roaring winds and high seas, the future and the past no longer exist.

Being trapped in a storm at sea is somewhat like being involved in a very prolonged accident. If you're good in that sort of situation, your senses are sharp and you feel calm and focused. If you're not good in crises and tend to panic or become hysterical, avoid a storm at sea. It's essentially a sustained emergency that will make you (and therefore your shipmates) miserable.

Amazingly, John and I were unafraid—maybe even a little excited. We felt very safe in the *Bossanova*. It seemed as though this was the weather she was built for, and though we were eager for the storm to end—no one wants to push his luck at sea—we felt a kind of cozy thrill when we had enough visibility to see the patches of fine white spray blowing off the surface of the ocean. It looked like snow being whipped across a wide field by a bitter wind.

Especially in periods of zero visibility, I was very conscious of how alone we were, trapped in our own world of weather. The air had become white with sea foam, whipped by a wind so loud it had overpowered every other sound and become its own eerie kind of quiet.

And just as suddenly as we had entered the storm, we exited. It was dusk and we were only minutes away from our destination. The waves died down, the wind petered out, the sky was streaked with a deep sherbet orange by the fading sun. As we passed through Rudee Inlet, the water was glassy, reflecting the beautiful sky like a mirror. Their very last slip waited for us, and as we pulled up to the dock, it was hard to believe what we'd just been through. The dockhand who greeted us shook his head in disbelief that we'd been out in *that* and made it back unscathed.

John found the bar while I took the dogs for a very long

walk around the suburbs of Virginia Beach. I was relieved to be safe and felt the tension draining from my muscles. I noticed something different in the air, too—in the way the twilight played across the grass, the bushes, the ditches of wet weeds. I had skipped spring that year; I'd gone directly from snowy Pennsylvania to hot Florida. It was the only time I could remember that I hadn't been around to rejoice when the first brave crocus peeked out from beneath a hat of snow. Now, I could suddenly smell summer and it reminded me of my childhood. In the northeast, summer is much more than another season—it is the sweet, soft answer to your frigid winter prayers.

On my way to the nearby restaurant, the skies opened up with another torrential downpour. I didn't mind a bit. John and I sat at the bar with good vodka and oysters Rockefeller and toasted the end of North Carolina, the beginning of the mid-Atlantic and, most of all, the mercy of Poseidon.

CHAPTER SEVEN

For whatever we lose (like a you or a me)
It's always ourselves we find in the sea.

—E. E. CUMMINGS

We left Rudee Inlet at 0730 hours the next morning. Looking back, it's clear that this day, Thursday, July 8, was a subtle turning point.

Although our first week had included three of our toughest days, our spirits had been up. We were excited to be out of the classroom and underway. We were resilient in the face of adversity. We were salty dogs, damn it.

But during the second week, a horrible malaise had descended upon the crew of the *Bossanova*. There was just no *joie* in our *joie de vivre*. We were tired. We felt hassled. John, Samba and Heck all got sick. The trip seemed endless. Of course, in the period between June 28 and July 5, we took three days off. It's no wonder that North Carolina seemed the size of Texas. This middle week was the worst part of our journey and very difficult for both of us, in different ways. I knew John had moments of regret for committing himself all the way to Sag Harbor, and I, in turn, felt guilty that I couldn't release him from his servitude.

But as we left Rudee Inlet bound for Chincoteague, the pall lifted. Just like that. There's nothing like escaping a huge storm unscathed to make you feel grateful for a dull day underway. And the whiff of summer I'd caught filled my nose like a bloodhound on the trail. We were getting closer—I could feel it now. Optimism returned to the *Bossanova* and perched on her bow like a seagull portending the first shout of "Land-ho!"

The *Bossanova* had an easy day with a familiar rhythm to it. I brought us out, then John took the helm while I made coffee and stepped outside to look at the passing sea. I still couldn't get enough of it—watching the ocean in all its thousand gradations of blue as the sturdy-looking bow of *my* little ship plowed it aside. It was my dream come true. When I had imagined a different life for myself, this is what I had pictured. And for once, it was every bit as satisfying as the fantasy had been. No, it was even better. I had never experienced such a sense of contentment, pure pleasure in the here and now, the deep desire to be nowhere other than exactly where I was. This was happiness.

FOR A TIME, HAPPINESS, too, had been Leslie. Tall, slender, blonde hair, great legs, big heart, blue blood, sexy and more fun than a barrel of chimps.

It wasn't love at first sight. We met at a party one night during my brief breakup with Laura. I ran into her again about a week later at a way-off-Broadway play. She called me the next day and left a message, asking me if I wanted to have dinner. I was already getting embroiled with Laura again and politely declined.

Several years later, my brother—who had never played matchmaker before—said, "I know somebody who might be perfect for you." By then, her name didn't even ring a bell.

But when she called me this time, I said yes. We had a great dinner at Clementine, where we seemed to do nothing but laugh. The next day, she hitched a ride to Connecticut with me—she was coincidentally spending the weekend with friends in my neck of the woods and she broke away at some point and came over to my house for a short visit. Leslie had left New York City to work in Los Angeles a few years before, and when she asked if I'd fly out and visit her the following weekend, I said yes.

I remember standing in front of the passenger terminal at LAX and seeing her pull up in her little Mercedes convertible (known among her friends as the Morgan Fairchild). As she got out of the car, dressed in a navy windowpane suit, her face lit up at the sight of me, and I thought, *Wow. Lucky me. This is the girl of my dreams.* "You are a sight for sore eyes," she remembers me saying, and that her heart fluttered at the words.

One of us flew cross-country every weekend but one that summer. We were giddily in love, head over heels. We wrote each other little notes, sent each other flowers and presents.

When we were together, we always seemed to be laughing. This was definitely the woman I'd been waiting my whole life to meet. She felt like home, and I felt completely married for the first time in my life.

In the fall, Leslie moved back to New York and into my apartment on Christopher Street—it was perhaps the first time she'd ever been below 14th Street and I'd check her Upper East Side skin for signs of hives each day, but she grew to love downtown very quickly. My apartment was charming but rundown, so Leslie insisted on renovating, even though it was a rental. We shared a deep appreciation for home design and we seemed to spend every waking moment poring over shelter magazines together, visiting showrooms and antique shops.

We both had new jobs. I was a content director for an Internet website that wasn't online yet. It was infuriatingly dull. Leslie had started a new job, too, and she was in a somewhat similar position, but making oodles of money.

So, there we were, living in a one-bedroom apartment that was a worksite, going off to jobs we didn't like—in retrospect, we should have had separate apartments for a while or perhaps moved into something that was new to both of us and ready to live in. Our life together was good but fraught with constant low-grade stressors.

In the spring we started house hunting. We looked at places in Connecticut, where I spent weekends, and places in the Hamptons, where she used to spend weekends. Though Connecticut was less expensive, it was quiet. That suited me fine, but Leslie was a very social person. In the end, I figured I could stay home if I wanted quiet, but if Leslie wanted more to do, she could find it in the Hamptons.

I remember when we found the plain Victorian farmhouse with two sweet guest cottages and a gorgeous pool in Bridge-hampton. We both loved it, though the house had a very dowdy interior: cheap brass lighting fixtures, heavily shel-lacked woodwork, ugly floral wallpaper. The previous owner was returning it to an authentic dismal-period look. But we could easily see our way past these cosmetic problems. We worked hard every weekend stripping wallpaper and paint-ing, and by the time summer came, we had an enchanting little compound.

But that's when the trouble started. We'd been under a lot of stress for the last nine months, between working at our bad jobs and living in a construction site, but we very rarely fought. We were kind to each other, gentle and forgiving when we did clash. Now, though, Leslie seemed to get further and further away from me. She was anxious much of the time, compiling lists of things we needed to get done. Our social calendar seemed constantly booked, and I was often bored by these obligations. Too many of these gatherings seemed to have an air of desperation about them: people wanting to see and be seen. It wasn't that these events were always fancy, but in this circle of the Hamptons, I never felt as if I could go to a summer barbecue in casual attire—"casual fabulous" was the de rigueur look. People competed, stealthily, to be the best dressed, the funniest, the richest, the sexiest. It was exhaust-ing and not very interesting to me. I should have just stayed home but I didn't.

Here is where a side of Leslie emerged that I didn't like much: she had always been the life of the party, but now I saw how much the need to be liked and popular fueled that need for fun. She could easily get caught up in this frantic social performance because it played to her insecurities. And yet, I knew that deep down Leslie felt the same way I did about these scenes. She didn't really have a snobby bone in her body, despite her upper-crust background and acquaintance with everyone who was anyone in New York. She liked nothing better than having a great conversation with an interesting person. She was deeply compassionate toward the less fortunate, and she went out of her way to help anyone she met, in any way she could. Sure, maybe a small part of that was fueled by the need to be liked, but not very much.

Later, we agreed that if I had moved to Los Angeles, we might have made it. I liked her friends there and they liked me. Surprisingly, her LA cohorts were all smart, funny, accomplished people, enduring the colossal superficiality of Hollywood so they could get ahead, without letting it transform them into "players." The New York circle seemed so much more interested in money, power and fashion than any of the other wonders that the Big Apple nurtures—they would have been better off in Los Angeles.

It was a wretched summer. My Internet company started collapsing and I lost my job. Leslie's situation at work had never resolved itself either, and she finally left with a hefty severance package. We fought daily and grew further and further apart. I couldn't find a way to get Leslie's attention anymore, to remind her of the bargain we had made to love each other forever. And I was petrified as I watched things unravel. She seemed anxious all the time, unwilling to draw boundaries between her friends and our relationship.

I had sold my house, bought one with her, shared an apartment in the city—my life was gone. It was *our* life now, yet I felt increasingly excluded from it in any meaningful way.

In September I moved out for a month, declaring as I went that I wanted to work things out, but that I didn't see us getting anywhere in our current situation when all we did was fight. Maybe some space would help us put things in perspective. Maybe we could decompress—have dinner twice a week and see if we could get back to solid ground again.

But Leslie wasn't interested and I never did go back. I spent most of October in my brother Hamilton's Bridgehampton summer rental (which he had taken through Thanksgiving), in bed, fully clothed, staring at a small black-and-white television. I don't know what was on because I couldn't see a thing. It was the darkest time of my life.

I was crushed by the failure of my relationship with Laura—I lost weight, I was very depressed—but it had been a long time coming and I knew it was right. Now, though, I felt like the most precious thing in my life had been jerked away from me. I felt stunned, traumatized, utterly disbelieving. Our breakup was a cruel mistake. I rationalized that it was messed up for Laura to give up so easily, but that didn't make me care any less about her or feel any better about my heartbreak.

And, stubborn as I am, some small but vital part of me kept pining for her, through every kind of ugly up and down.

After we broke up, we spent a year hanging around together, toying with reconciliation but never making it happen. When I finally started to date, Leslie had a complete meltdown and convinced me to try again with her for a week. It was a week that I spent, literally, a thousand miles away on a business trip, feeling suspicious and resentful of her timing, because it had taken me so much will power to start to move on. I still loved her, but I didn't trust her motivations and I continued my new romance. A few months later, Leslie started to see someone, and then it was my turn for a complete collapse.

I know—it all sounds very screwed up. But what's at the base of it all, our friendship, is not screwed up. We have been each other's favorite person on the planet, and no amount of icky drama (which I will spare you) has managed to ruin that.

I used to feel that someday we might get back together. Over the last few years Leslie has grown so much—she's much more secure, less eager to please, capable of intimacy with both friends and lovers. I'm happy for her and proud of her, but it's also painful for me to see: she's becoming someone that I *might* have spent my life with. We always agree: at a different time maybe we could have been the right people for each other.

And for a long while, I felt positively haunted by my love for her. Stalked, even. There's no doubt it was a big part of why I ran away to sea.

Now I was doing something with my life that Leslie would never have done—and I wouldn't have done it if I'd been with her. Leslie hated boats; she always got seasick. But I was feeling completely blissful while underway aboard the *Bossanova,* and I realized one day that I had stopped pining. I just *was.* That sensation was unbelievably freeing. I accepted, at last, that I would always love Leslie, even though we were over. I began to believe that there were other people I would one day love as much, but differently. I got comfortable with my grief and acknowledged that I might always have to lug it around, but that its weight made me stronger. I even suspected that one day, it would feel as light as a feather.

And it was this sense of liberation that made me understand what my journey was really all about. Someone once suggested that I was redefining myself, but that wasn't true. In fact, I was *undefining* myself. It was as though I'd made a list of "Everything I know about me" and was just erasing each item, one by one.

What would I do when the page was blank?

WE MADE CHINCOTEAGUE INLET in the late afternoon. The ride up the channel to the town dock seemed endless. We tied up behind a long, red, slightly rusty fishing vessel, rigged with bright lights for working on deck at night. A policeman on a mountain bike came to the boat to collect the dockage fee: $40 for a piling to tie to—there was no bathroom, no laundry, no fuel facility, no other boats, no real slip, absolutely nothing. The small town looked a little worn-down, quieter than you'd expect given the endless number of store windows that hawked merchandise celebrating the swimming Chincoteague ponies. It had that "If we build the T-shirt stores, they will come" feeling, a kind of desperate optimism in the face of failure. Because of this, its authentic charm hadn't yet been subsumed by fake charm and I liked it there.

We had a fun meal that night at a bar on the water. Two old-timers sat next to us, drinking Amstel Light and reviewing recent movies. They looked like doppelgangers for Walter Matthau and Jack Lemmon in *Dirty Old Men*, but they talked Hollywood like they were at the Ivy with Bob Evans.

The bartender was busy with the waitresses' orders, so we sat for at least twenty minutes before he remembered the parched people at the bar. I made the mistake of risking a margarita on the rocks, then watched with horror as he used Rose's lime juice.

When the place quieted down a bit John struck up a conversation with our barkeep. He was good-looking in a cheesy kind of way—tight jeans that were faded beyond the point of fashionable, a surfer necklace and a blue short-sleeve shirt that was no doubt carefully chosen to play up his eyes. Since I expected him to pull a comb out of his back pocket at any moment and do a little primping, I wasn't surprised when he confessed his goal of moving to New York City to become (gasp!) an actor. I resisted the urge to ask him to try Method-acting a good bartender. (I'm always a wee bit irritable after the senseless slaughter of an innocent margarita.)

We were off bright and early the next morning. Shortly after we reentered the Atlantic, we heard a communication from a 90-foot fishing vessel that was 30 miles offshore. They reported 54 inches of water in their engine room and rising. A high-speed Coast Guard rigid inflatable boat went flying by us, and we heard a navy warship announce it was altering course to assist. A Coast Guard helicopter was also dispatched to drop a pump and possibly remove the crew.

Later that day, we heard another Coast Guard call seeking information on a vessel that disappeared after issuing a Mayday. This kind of message went out several times over the course of our voyage, and it was always chilling to wonder what had happened. A Mayday is the most serious of distress calls—even the fishing vessel on its way to sinking used the urgent but less dire Pan-Pan signal (pronounced "pon-pon"), as did the pleasure boat off the Frying Pan Shoals. A Mayday (from the French, *m'aider*, "help me") is to be used only when your life is at immediate risk. So a vessel that issued one and then disappeared was obvious cause for concern.

For the *Bossanova*, though, it was another banner day. After a week in North Carolina, we felt like veritable speed demons as we left Virginia behind and breezed past Maryland and Delaware on our way to Cape May at speeds reaching close to 10 miles per hour. *Wwwwwhhhheeeeeeeee!* This was one of the loveliest days of our trip. It was clear and sunny and we were treated to a constantly changing display of aquatic life.

Midmorning, we saw two giant sea turtles mating. Several miles off the coast, there they were, gettin' it on in the middle of nowhere. Get a room, I shouted to them.

We saw a solitary shark, too. The casual but relentless back and forth of its dark fin against the sea made me shiver in sympathy for its prey. It was good to be standing in 30 tons of steel.

There were countless stingrays and millions of jellyfish. Their small, nearly invisible bodies created an enormous gelatinous river that coursed through the ocean for miles.

We'd seen dolphins the whole way up the East Coast, but they were most plentiful and most playful in the South. They frolicked in Florida, leaping out of the water in small, synchronized groups. The farther north we went, the less abundant and more lethargic they appeared, their arched backs clearing the surface but not much else. Still, they were always exciting to see and we kept a close lookout for them wherever we went. They felt lucky to me—and months later I read that dolphins swimming with a ship are, in fact, always considered a sign of good luck, according to sailors' lore. (On the other hand, women onboard a ship supposedly make the sea angry. But a naked woman on board will calm the sea, so I supposed that my excellent personal hygiene was evening things out.)

Today, we were treated to a big dolphin revival. They were probably drawn by the slick of transparent delectables we motored through. At one point, we counted as many as eighteen dolphins together.

As we approached the unlikely mecca that New Jersey had become, a quartet of dolphins peeled away from the larger pod, doubled back and got directly in front of us. It was thrilling to stand 10 feet above them on the bow and watch their subtle choreography, their perfect calibration, just beneath the water and a yard before my boat. There was something so beautiful and friendly in their spontaneous escort that it made my eyes water with happiness. The dolphins ran before us for about three minutes and then veered back toward their pod. It was hard not to feel we'd been honored.

At Cape May, we pulled into a large marina with a huge restaurant and small tiki bar. It was Friday night and the joint was jumping. I joined John for a drink, then left him chatting with a few guys and went back to the boat for a cold dinner with the boys and Anna Karenina.

I was finally making some progress with old Anna K., after a very slow start. Although I had read Tolstoy's masterpiece before, I couldn't remember a thing about it so I thought it would be a meaty choice for a long trip. One of the few advantages of a lousy memory is that you can reread all your favorite books, rewatch all your favorite movies and enjoy them as though it's the very first time. In theory, anyway. What I discovered is that my tastes had changed a lot in the twenty-eight years since I had last read *Anna Karenina.* Why a prepubescent girl would find the politics of prerevolutionary Russia more interesting (or at least less dull) than a 40-year-old woman disturbed me. Had I become dumber with time? But as I forced myself onward, I was gripped by the intricate character portraits Tolstoy drew and by his understanding of human nature. Maybe this was what I had liked about the book the first time.

The last time I'd read the book was during "The Year of the Russians." I was 13, and we were living in Ireland at the time, at Walker's Lodge in Sligo. I was a reading maniac. We all were. There wasn't much else to do after we'd finished dinner in the kitchen, the only warm room, which was heated by a big coal range. Sometimes Dad would quiz us on current events or literature. Or we'd listen to Radio Luxembourg count down the top twenty hits and send out dedications while we did our homework at the kitchen table. But bed, with its promise of warmth and privacy, soon beckoned. Armed with a hot-water bottle and a kerosene lantern or candle, off we'd rush to our imaginary worlds. The wind and rain rattled the windowpanes and made the low light flicker on the page. The smell of a struck match still makes me think of those nights in Ireland.

I'm not sure why I chose to obsess over the Russians that year, but I've always considered it a miracle that I'm neither blind nor severely medicated. Maybe nightly sleigh rides through the bitter cold of St. Petersburg and hard time in the permafrost gulag helped make cold, damp Sligo a cheerier reality by comparison. Maybe this was the beginning of my habit of reckless optimism and deep denial.

The next morning, John and I chugged up the Jersey shore, past Avalon, Sea Isle City, Ocean City and Atlantic City. We stopped for the night at Point Pleasant Beach, just south of Spring Lake. We had run as far as we could before dark, and Garden State Marina was the only place we could find with a slip for us. It was a summer Saturday night on the Jersey shore, so that made sense. Sometimes we completely lost track of what day it was while we were underway.

The marina was closed when we arrived, but we'd already paid and been told which slip to use. Unfortunately, it was a slip for a 60-foot boat and the pilings were spaced far apart; a stiff breeze pushing us to port made it difficult to tie up. When we finished, we were treated to a delightful welcoming committee. A swarthy guy with a big scowl on his face came down and said, "Does Richard know you're docking here? Did he say you could dock here?"

"Hi. How you doing?" I asked pointedly. What a rude jerk. "Yes, he certainly does." What I wanted to say was, "We're paying $120 to stay in this crap hole, so why don't you at least be civil, if friendly is too much for you to manage?"

I was in a bad mood as I walked the dogs at dusk around the small blacktop parking area that Garden State called a marina. John had gone to find us a dinner spot, but I wasn't hopeful. I could see a line of fast-food joints on either side of busy Route 35 South. But I should have had more faith—not in Point Pleasant, New Jersey, but in John. He called me with elaborate directions to some place called Stretch's, about ten minutes away. I had to walk across the highway, down several blocks to a restaurant called Tesauro's, cut through their parking lot and keep walking down a road near the water. John warned me that it was a dark and kind of a dubious-looking neighborhood, but when I had passed the easy chair someone had dragged to the curb, I'd know I was almost there.

When I arrived, John was in seventh heaven. He was halfway through a dish of giambotta, a dish he ate frequently back in Chicago. "Oh, Mare. You gotta try this. It is out of this world." I had a forkful—it was delicious and, except for the addition of sausages, a lot like ratatouille. I knew if I told John that his beloved Chicago dish was reminiscent of one of my beloved French dishes, it would just ruin his appetite. The French were one of our verboten subjects—John would flash me a dirty look if I had a glass of red wine.

Stretch's was a great find. It was an unpretentious place with a great menu and a homey atmosphere. Best of all, there was a jazz duet playing—one guy on keyboards and one on guitar. I asked them if they could play "Wave" and they looked thrilled that someone was actually listening. I got "Wave" and then I got two or three other Brazilian classics without asking. Point Pleasant beach was saved. I'd even go back in a car, if I had to.

John was happy, too. He hadn't eaten this well since his last trip home. The Windy City has never had a finer ambassador than John, who makes it sound like an exotic gourmet paradise. He was always getting misty-eyed about Italian beef, or giardiniara. During the long, uneventful hours we spent underway, John would often tell me stories about his friends back home and the places they hung out. By the time we made it to Sag Harbor, I knew I'd miss John and many of his favorite dishes, not to mention Skychair Bob, Red, Voges and all the rest of his gang of "knuckleheads." (This endearment of John's was both a dig and a huge sign of affection. When I overheard John one morning greeting Heck and Samba with a "Good morning, you two knuckleheads," I knew they were pals.)

It continued to amaze me that John and I got along as wonderfully as we did. We joked about sending our Chapman colleagues an e-mail to let them know that after three weeks together nonstop, I had become a Chicago Cubs fan and ardent Republican and John was listening to NPR and reading *Anna Karenina*.

"Nah, Mare," he said. "That's going a little too far. No one would *ever* believe I'd listen to NPR."

On Sunday morning, we had yet another lovely summer day before us. I had long harbored a fantasy of a triumphant New York City arrival, circling Manhattan in style, maybe inviting some people down to the dock to have cocktails aboard my little ship. But time was running out. We also had to scrap a plan of cutting through New York Harbor to Hell's Gate and then motoring up Long Island Sound. We knew the currents at Hell's Gate could be tricky and our passage should be timed to go with the tide. That would mean waiting around in the East River. Since John had a flight out of LaGuardia early Tuesday morning, it looked like we were going to be cutting it close no matter which route we chose.

I studied the charts. It seemed it would be both faster and less stressful to stay on the outside of Long Island and cut through the Shinnecock Canal. My own personal Circle Line tour was going to have to wait for another day.

After we'd run about 15 miles up the New Jersey coast, we decided to cut straight across to Long Island. It would put us roughly 15 miles offshore at one point—farther than we liked—but we'd be able to completely bypass busy New York Harbor that way, and at least make Fire Island before the end of the day.

One of the great moments of my life was standing at the helm of the *Bossanova* looking at Manhattan, home, just 10 miles off in the distance. The city looked as majestic as ever, though it would always be different without the twin towers. Now, looking at the skyline was like seeing a smile that still dazzled even without front teeth. *Love is not love which alters when it alteration finds.*

Today, looking across the water at New York, I was dazzled by its verticality. (I once heard that if everyone in New York spilled out onto the sidewalks at the exact same time, the streets would be forty-three people deep!) I imagined the millions of different lives happening in that little space: people eating brunch, making love, watching TV, paying bills, walking dogs, fighting, jogging, reading the *New York Times*, shopping, biking, singing, dying—all at once, right over there. If you could take a cross section of the buildings and look down at it all, it would be exactly like an ant farm: teeming with intricate and pointless industry to a distant viewer, yet full of meaning and purpose to the ant.

It was a Sunday morning in New York, and I could imagine my friends Julie and Adam and 9-year-old Jackson in their apartment right now as vividly as if I was standing there with them. It felt good after the long trip, even if it was an illusion.

I like to close my eyes and vividly picture myself in different places. I developed this habit of mentally projecting my physical self in Brazil when I was a 16-year-old exchange student. Whenever I felt really homesick, I would close my eyes and imagine what was happening in our kitchen back in upstate New York. I wasn't particularly attached to the house, where I had only lived for a year, and I didn't really like upstate New York. But despite my very nomadic childhood, I'd never been away from my family. And I suppose that I was more fiercely attached to them as a result of all that moving. At home, *All Things Considered* would be on the radio, Mom and Dad would be having cocktail hour, leaning against the old refectory table while they fixed dinner. Sometimes I smelled hamburger and onions simmering in olive oil. And I would imagine myself touching that table, stirring the food in the pan, helping myself to a piece of cheese and a cracker, like a ghost wandering through the present. But this wasn't mere reminiscing, a nostalgic savoring of the details of home. It's hard to explain, but on these visits, I forced myself to understand that there was only a fine line between actually being there and being there in spirit. I would remind myself that I could actually be standing in that kitchen in less than 24 hours if necessary, and if I did that, it would then all seem so ordinary, so much as I pictured it, that there'd be no point actually *being* there. Visiting home in my imagination helped relieve the pressure of separation anxiety and let me feel like I could almost be in both places at once. It's a trick that comes in very handy now, whenever I'm away from the boat or landlocked for any amount of time.

I decided to give Adam and Julie a call from my cell, which was working. Un-friggin' believable. I didn't know whether to be glad or annoyed. Half the time you couldn't get a decent signal standing on top of a cell tower, and here I was 15 miles off the coast.

"Wooooooo-hoooooooooo," I greeted them.

"Wooooooo-hooooooo-hooooo," they whooped back on speaker phone. "Where are you now?" they asked. They'd been checking in with me every few days to monitor my progress up the East Coast.

"Well, I called to tell you I'm passing New York right now and waving at you like a crazy person. We should make it to Sag Harbor by tomorrow night."

There was some more wooo-hoooing, with promises to see each other the next weekend, and then I hung up.

As New York disappeared behind us, we realized we were making much better time than we'd anticipated and altered course slightly to aim farther up the coast of Long Island. Later, when we were sure we could make it, we altered course again and continued on to Shinnecock.

Shinnecock was so close to Sag Harbor by car that it was hard not to feel we were practically home when we tied up for the evening. We were now in a neighborhood that a Los Angeles realtor would call "Hamptons-adjacent," and if we hadn't realized it before, we certainly did when dockhands in white uniforms arrived to help with the lines. The sun was just starting to sink, and the restaurant above the docks was jammed with women in Capri pants and big sunglasses and tanned men in golf shirts with booming voices. The tables were littered with cosmopolitans and Judith Leiber handbags, and the restaurant overflow spilled onto the deck, shouting over the sounds of a steel drum band.

We decided not to battle the crowds and found our way to the deserted inner bar. We toasted an amazing trip and our most successful day, and ordered the best crab cakes I've ever had. I tried to pry the recipe out of the chef when he came out to the bar, but he could not be bribed.

John and I were now becoming very nostalgic about the end of the trip. We even looked back at the purgatory of the middle week with fondness. What would our voyage have been without at least a little bit of suffering—and we'd gotten off so easily! I had to admit, I was ecstatic about the *Bos-*

sanova. Not only had she seen us through some really rough weather, but she'd made her very first trip with me, over a thousand miles, without a single mechanical problem. (I didn't count the two batteries we'd replaced partway up—that was more of a maintenance issue.) Even though we weren't home yet, we were now close enough. On the off chance that we suffered some massive failure the next day, John would still make his flight and I was only a phone call away from a bed at a friend's.

In the morning, we pushed off a little bit later than usual, around 0900 hours. The skies were gray, the clouds looked stuffed with rain. We motored through the approach to the Shinnecock Canal and wondered what it would be like. As we approached the lock, I looked up at Route 27, the highway that passed above it. Over many years of driving that road to the Hamptons, I'd developed a little ritual about looking to the left, right there, toward Great Peconic Bay. It was always the first glimpse of water after driving out of Manhattan, and seeing the boats arrayed at the marina below made me feel happy and excited. I had never realized that this was the Shinnecock Canal, that there was a lock here, or that I would one day bring my own boat beneath the highway and out into the bay.

Since we'd taken the Atlantic route, we'd had no need to use the systems of locks that link many inland waterways. They're designed to convey boats from the level of one body of water to a connecting body of water at a different level. I'd heard they could be really harrowing but had no idea what to expect. Up ahead, we saw that traffic was divided into two lanes. To port were Atlantic-bound vessels, to starboard, boats headed for Great Peconic Bay, Gardiner's Bay and Long Island Sound. A Boston whaler coming through to the Atlantic side fishtailed wildly as it exited the lock. The skipper looked like he'd lost all control as he veered toward the retaining wall, then fishtailed back away from it and recovered the middle of the channel.

"What the hell happened to him?" John asked.

A green light ahead of us indicated that the lock was open and we would not have to wait. Excellent! But as we pushed toward the gates of the lock, we felt the amazing force of the current rushing against us. I throttled up from the gentle 1,000 rpms we'd been running at as we approached. We were being pushed back. I kept pushing the throttle up, past 2,000 now, and we were still being forced ever so slightly back. As we hit 2,500 the engine's growl got deeper, throatier, and the boat finally stood still. We were trapped squarely between two concrete walls 15 feet away from either side of the *Bossanova*, with another boat not far behind us and we still weren't making any headway.

"Oh, my god, John. I can't believe this. I'm not sure we can make it." I was perhaps more panicked than I had been at any other moment of our trip. Turning around would be very difficult against this current. At the same time, I wasn't sure I could control the boat if we had to back out—and the vessel waiting behind us added to my nervousness. I did not want to tangle with it. I gave the engine one more blast of throttle, pushing her to her limits, and—hallelujah—we surged forward.

Of course, now we knew why the boat exiting the lock toward the Atlantic had been swinging around—he had the force of the current on his stern pushing him out. This close call at the locks was something we hadn't anticipated, but once we were through, we were a little giddy. We knew it was likely to be the last adrenaline rush of our journey, and we were almost glad for that final taste of challenge.

As we cleared the canal and chugged around toward Sag Harbor, I watched the coast for landmarks. But everything looked different from out here. I wouldn't have had a clue where I was without a nautical chart.

We rounded the point into Sag Harbor around noon. I can't possibly explain how happy I was, how exhilarating it felt to tie up at the town dock and climb onto a pier I'd walked down a hundred times before. One summer I had belonged to the gym near the wharf, and I'd never been more diligent about working out. The view of the docks, stacked with superyachts, as well as the view of the harbor's more humble moored fleet, always filled me with a sense of peace. That was only a few years ago, yet I had never once looked out at those boats and thought I might actually own one, live on one, pilot it over a thousand miles through the Atlantic.

The major accomplishment of what we'd done had really escaped me until this moment, but now it was sinking in. I had brought my boat all the way up from Florida in under three weeks, on $600 worth of diesel, with no breakdowns and no disasters. We had stepped aboard in Stuart, Florida, and here we were, disembarking in Sag Harbor, New York. The trip was over. We'd done it.

John and I went immediately to Dockside to celebrate. We had lunch and toasted an amazing journey. John said, "Mare, I gotta tell you. It was a great trip. It couldn't have been better. I actually feel sad that it's over. And I've gotta admit, the ole *Bossanova* is a hell of a boat. Thank you for asking me to come with you. Really, it has been one of the best experiences I've ever had."

I hardly knew what to say.

"Hey, I don't know why you're thanking me, John. I never could have done it without you. Truly. And even if I had found someone else to come, it wouldn't have been the same. There's no one I'd rather have done it with." And much to my surprise, that was absolutely true.

I was too wound up to sit at the bar and drink all day, so after lunch, I left John and walked around town. I knew I had a ridiculous grin on my face and my feet felt like they were hovering above the sidewalks. I was giddy, positively high. I simply could not believe that I was in this town, which I thought of as my second home, and that I had made it here in my own boat.

Let's face it—it's been done a couple hundred thousand times before. It was a lot bolder than coming up the ICW, as most people do, but it wasn't nearly as daring as climbing Everest or sailing solo around the world.

But the joy I felt today had nothing to do with risk-taking or daring. It especially had nothing to do with what anybody else thought of my trip. I had finally done something that was intensely meaningful to me. More meaningful than good grades, scholarships, speedy promotions, bestsellers—

everything else I had done *right* in my life, all my other "accomplishments." I had gone from knowing nothing six months ago to coming up the East Coast through the Atlantic as captain of my own boat, and it was the greatest thing I'd ever done. No doubt about it.

The next morning, John and I got up early to drive to the airport. It was pouring rain, but we had a head start, so I wasn't worried about missing the flight. Silly me. We were trapped for what seemed like an eternity on the Long Island Expressway, in the 30s exits that have been under construction since the dawn of man and probably still will be when the last star fades from the sky. The minutes ticked by, and it began to seem like there was absolutely no way John was going to make his flight. We laughed about how funny it was to come all this way and then screw up something this simple. I dropped John at LaGuardia feeling lousy that he'd probably have to wait around for the next flight.

Forty minutes later, he called me from his cell. Apparently Poseidon was still watching over us. The rain that cascaded from the heavens had delayed all departures. John was sitting on the plane, preparing for takeoff.

The trip was over.

CHAPTER EIGHT

Fortune brings in some boats that are not steered.
—WILLIAM SHAKESPEARE

Well, I *thought* the trip was over. There was one more leg from Sag Harbor to Maine that I planned for the end of summer, and there was still the unknown experience of truly living aboard a boat waiting ahead for me. But I imagined that the biggest surprises of my voyage were behind me. But I failed to consider the real meaning of my journey. There were still one or two things I would learn about myself.

After I'd enjoyed a couple of luxurious days at the expensive town dock, I relocated the *Bossanova* to a cove, just under and beyond the bridge to North Haven. It was free, and, better than that, it was a quiet, sheltered spot away from the flashy superyachts that thronged the town's waterfront.

While I was tied up in town, I'd made friends with Matt, the guy who ran the Sag Harbor launch, so any time I needed to get to land I'd call him up on the VHF and he'd make a quick pickup for $3. He was also great about bringing me back after hours on the few occasions I stayed ashore for dinner. I'd give him a call and he'd come down and meet me at the docks—I gave him a big tip on these trips. Of course, I'm sure that a big gratuity from the owner of the *Bossanova* was not quite the same as a big gratuity from the owner of a superyacht—although, come to think of it, you never know. It's not the size of the boat that matters.

Naturally, I had my own dinghy aboard the *Bossanova*, but the cowl that covered its outboard engine had broken off. It smashed against the deck when we tangled with that first storm in Georgia, and the next day it slid into Charleston Harbor when we almost capsized. I'd have to track down a new one, but in the meantime, the town launch was much faster, and it was an easy way to bring the dogs to shore for exercise. The three of us in my little 9-foot inflatable would have been a little bit fur-raising.

Life on the hook, as salts call anchoring out, was fantastic. I had all the amenities of a rich resort town nearby with none of the hassles or expenses. I'd bring groceries back and grill, sit and watch the sunset with a glass of wine. It was wonderful. Every time I was in the water taxi and rounded the marker just past the bridge, I looked eagerly for a first glimpse of the *Bossanova*. There she was, sitting majestically in the cove, looking like an ex-navy boat and a fish out of water in the polished Hamptons.

A few days after I'd anchored in the cove, I noticed the VHF signal petering out to nothing. Although I ran the engine for an hour every morning and evening, the batteries just weren't holding a charge for very long. I knew I was going to have to figure something else out. Though I was loath to spend any money on a berth when it was so nice at anchor, I knew it would be easier on all of us (that is, me and the salty dogs) if we could plug into shore power and get on and off without a VHF call and a 20-minute wait. The real difficulty would be finding a slip. I'd already phoned or visited all the marinas, and there was nothing available for the rest of the season at any price, never mind within my laughable budget.

One evening, when I'd decided to spend a quiet night on the boat with the boys, I was sitting in the salon reading. Almost all of the power was off to conserve energy, and it was getting so dark that I struggled to see the page. Suddenly, I smelled something. Smoke? I jumped up and ran below, throwing on the engine room lights as I entered. How could there be smoke coming from down here? I thought I must be imagining it—the engine wasn't running, almost nothing was on. I cast a glance at the starboard side of the engine, which looked fine, and circled quickly to the port side. Oh my god—fire!

Fire on a boat, even a steel boat, is every captain's worst nightmare. More boats are lost each year to fire than they are to the sea. It seems odd, since fire's natural enemy should be an aquatic environment, but a boat has a hot engine, a large fuel supply and an electrical system—all of which are constantly exposed to the unrelenting corrosiveness of the marine atmosphere.

I grabbed a fire extinguisher off the bulkhead, pulled the ring, aimed and squeezed. The fire sputtered but didn't go out. I took another extinguisher down and tried again, shooting carefully for the base of the flame. Same results. This wasn't working. Trying not to panic, I turned both battery switches in the engine room to OFF, then ran up to the pilot-house and threw the main circuit breakers—just to be safe. I filled a small bucket with water and hightailed it back down to the engine room. I heaved the water and watched as the flames instantly died a smoky death.

I breathed an enormous sigh of relief and gratitude. Thank god I happened to be at home that evening, otherwise, the dogs would have been alone and the fire would have gone unchecked. Disaster. And I couldn't quite believe that I'd summoned the presence of mind to throw the battery switches and circuit breakers off before I tossed the water on the engine. It can be a deadly mistake to use water on an engine fire. If the fire is electrical in nature, it's a great way to be electrocuted. I was also really glad this had happened here in Sag Harbor, instead of when we were in the middle of nowhere on our way up.

After sitting in the dark salon for a few minutes listening to my heart rate return to normal, I took a flashlight and went below again to see the damage. It looked worse than it was, thanks to the crusty yellow-white powder left by the extinguishers. After I brushed the residue off, I could see that the burned area was quite small and the damage minor.

I thought back to those dull hours at Chapman, learning about engines. What was this, I wondered, and why did it catch fire? I traced a couple of wires, several of which had melted, and pieced together an idea. I went back to my stateroom, pulled down Nigel Calder's excellent book on diesel engines and found a diagram that I needed to confirm my suspicions. Yup. The wires connecting the solenoid starter to the alternator had somehow caught fire. Those wires would obviously have to be replaced, and it looked like a portion of the solenoid, the cap on top, was melted.

I have never harbored any illusions about my mechanical ability. While I wished I could learn to repair my own engine, I was realistic about the likelihood of that happening. I decided early on that I'd be happy if I could articulately describe to a mechanic what was wrong. But even my victorious identification of the problem didn't clarify what had caused this fire. The alternator wasn't running, the engine was off, most of the power on the boat was shut down...

In the morning, when Matt picked me up so I could do a few errands, I asked him about local boatyards. He recommended a place called Ship Ashore that was just a stone's throw from where I was anchored. I should ask for Rick. I called when I got home that afternoon, but Rick said they were closing at 4:00 p.m. and were pretty busy until then. They'd have to come on Monday.

Monday! That was three days away—and since I couldn't start my engine, the batteries were going to run all the way down. I would be completely without power for most of the weekend. The lack of amenities didn't bother me much, but I didn't like the idea of floating in 30 tons of steel without a way to maneuver. What if I dragged my anchor, for instance?

Although I had described the problem to Ship Ashore, it was still reasonable to assume they'd show up on Monday and tell me they had to order a new solenoid—it was unlikely they would have one in stock. Then I'd be without power for several more days while we waited for it to arrive. All right—there was one thing I could try to do: find and order the correct solenoid. Maybe I could even install it myself before Monday.

There's a guy named Bob Smith who is something of a legend in the trawler world. Rumor has it he can disassemble and reassemble a Ford diesel engine in less than 30 minutes. I dug out a back issue of *PassageMaker* magazine and found his small advertisement. I called, described the problem and got the kind of great service that's become practically extinct now. Bob wasn't there, but some patient soul coached me through finding the part number on the solenoid. It wasn't easy because it was located in an awkward spot and had been partially burnt. But several phone calls back and forth, and my replacement solenoid had been located and shipped for overnight delivery. I felt very proud of myself.

The next afternoon, I got down underneath the engine with a series of wrenches, trying to be very butch about it. I may even have let my jeans hang down a bit, flashing my "Deer Isle smile," as a friend of mine calls it. I was determined to give this my best shot.

It didn't look very complicated. Five minutes, three scraped knuckles and at least fifteen expletives later, I gave up and decided to drink an alcoholic beverage with my pinky extended. It's healthy to be able to acknowledge both one's limitations and one's gifts.

On Monday, a guy with a tool kit showed up. He had longish blonde hair, a baseball cap with an unbent brim, a deep tan and a Long Island accent. He looked like a typical guy who messes about with boats, but his name was Moishe. I loved that. Moishe and I chatted while he effortlessly removed the solenoid.

We'd already met. He had come by in his dinghy a few days before and circled my boat admiringly. He'd asked me a lot of questions, and we'd talked about the former Russian pilot boat he lived aboard.

I'd always been a little reclusive by nature. However, something about life aboard the *Bossanova* changed me. Now, whenever somebody admired my boat, I asked him if he wanted to come aboard and have a look. It didn't matter what I was doing—in the middle of dinner, reading a book, just out of the shower. It started as pride in my vessel and an affinity for anyone who liked her. But it grew into a general openness to new people. Most boaters are friendly and good-natured. They are *invested* in their relaxation. After all, you have to be pretty comfortable with yourself to spend all that time offshore doing nothing but looking at the changing colors of the sea.

I'm sure, too, that all the generous cheers I'd received for my docking efforts had slowly melted my defenses. There's a great sense of camaraderie among boaters and it was infectious. One day I had not one but *two* conversations with men I had just met, who asked me all about the *Bossanova* and concluded our conversations by saying, "You're my new idol." I don't know about your life, but idolatry doesn't come along all that often in mine: twice in one day ran the risk of ballooning my ego past the bursting point. And the fact that these were admiring *men* was especially heady.

I think men were often surprised that I could handle what was really a small steel ship, that I had the courage to bring her up through the Atlantic and that I had the "balls" to get a gritty workboat instead of a cute fiberglass replica. The *Bossanova* intrigued people everywhere she went, but she also gave me a lot of sea cred.

But it wasn't just my tickled ego that was feeling good—I was happier and more relaxed than I'd ever been.

Several times over the summer when I was out with friends, someone I hadn't seen in a year would say, "Wow! You look so different." At first I assumed it was just the extra Pennsylvania pounds I had lost, but whenever I pointed this out, they'd say, "No, no. It's your face. You just look really happy, peaceful." I was, of course, but I couldn't help wondering how I must have looked before. Did I grimace? Scowl? Sneer?

After Moishe and I had chatted for a while, I went about my business and left him to his work. He made a trip back to the marina for a tool he needed, but after about an hour and a half, he emerged from the engine room to say I should be all set. The wires were replaced, the new solenoid was attached. Now was the moment of truth. We set the battery switches back on, I went up to the pilothouse, turned the key and pressed the starter button. The good old *Bossanova* coughed deeply and then chugged back to life. God, I loved my little ship.

Moishe's guess at what had caused the fire was only

slightly more scientific than mine. Some cables that were held aloft near the alternator had slid off a brace, and two live wires had somehow jiggled against each other. In other words, it was just a freak accident.

I was grateful more damage hadn't been done, but I dreaded the boatyard's bill. It's common knowledge that there is no better way to go bankrupt than boat ownership, though I comforted myself with the fact that I was pretty close to bankrupt already, so it wouldn't be the same agonizing fall that it would be for, say, Donald Trump. The solenoid had already set me back about $200. I wondered what Moishe's hourly rate was and whether the trip to the marina and back had been on the clock. In the end, the total was about $150— much better than I'd feared. And Rick, the owner of the marina, wound up offering me a great deal on a slip that had just opened up. I am a lucky, lucky person.

My first week back in the New York area, I scheduled lunch with an old publishing colleague so I could give him a full report on my trip. We were meeting at a power lunch spot for media types. I got there first and sat down at my friend's regular table. Parading by me was a veritable who's who of New York. I was never really a part of this scene, even in my publishing heyday. I'd have a lunch somewhere like this when I had to pull out all the stops for a celebrity author, and back then, I'd just be amused by it all.

Today, as I watched former Mayor Dinkins go by and Regis Philbin come in and Liz Smith be seated near the network honchos and other media elites less familiar to the public, I felt...tired. It would be too extreme to say my soul curled into the fetal position, but it definitely pulled the covers up over its head. I hated being there. Something had happened. Some last slender thread that had kept me a part of the buzzing Manhattan media scene had snapped. I couldn't see this venue as anything but a hotbed of vanity, and while I'm sure that the food was delicious and everybody there was very talented at what they did, I felt turned off by the palpable conviction that there was a lot of important stuff happening here. I missed the dolphins.

Clearly, I was finished as a book editor. My old life, even if it had been available to me, was not something I could comfortably wear anymore. I brightened when I realized that nobody was going to make me go back. When lunch was over, I'd get in my car and return to the boat. And tomorrow, I'd get up and do some freelance writing or pick up some of the decorative-painting work I sometimes did, for a sliver of what I used to earn but twice the satisfaction. By the time my friend arrived, I had come to terms with my own willful underachievement and gleefully embraced my lack of ambition. Arrghhhh, matey, I greeted him. At least in my own mind.

A few nights later, before I moved into the slip at the marina, I had planned to meet my friend Manuel for dinner at the Beacon restaurant, which overlooks the cove where I was anchored. I had spent the day painting with my friend Jay, and we knocked off later than I'd expected. I decided not to go back to the boat but to shower at Jay's and dash to T.J. Maxx for something to wear that was casual but not splattered with paint. I had lost weight in the last two months and needed some new pants anyway. After a mad race to get there on time, I arrived breathlessly in blue jeans and a white man's shirt. It wasn't chic but it was me.

Manuel, of course, was looking as elegant as ever in white jeans with a silk shirt and a cashmere sweater tied over his shoulders. We kissed hello in the bar, and he said he'd been admiring my boat while he waited. "You moved it, didn't you?" he asked.

"Yes, but I'm amazed you can tell." I'd motored a hundred yards away and reanchored in compliance with local laws that prohibited staying in one spot longer than seventy-eight hours. "You're very observant," I remarked, turning to behold my baby...*who was nowhere in sight.* My eyes scanned the water and there she was—several hundred yards from where she'd been, and up against the North Haven shoreline!

"Oh my god, Manuel. That is *not* where I put her. She's dragged her anchor—I've got to go. I'm sorry."

Three minutes later I was in the launch with Matt and we were flying out toward the *Bossanova*'s new resting spot. Matt had heard on the VHF that the harbormaster was also on his way. This was very embarrassing, and I was hoping to get aboard and take control of the situation before he had to. I couldn't believe this—how had it happened? When I'd reset my anchor, I'd tugged on it very hard and it was definitely holding. Also, I'd been sitting securely in the same spot for two days. Why would the anchor break loose now?

As we approached, I saw an older, very Greenwich-looking couple in a rowboat. They had a springer spaniel with them. The man had on a kelly green cable-knit sweater; his wife was wearing a matching one in yellow. She half-stood and called out to us.

"There are dogs on that boat," Muffy yelled in an apparent state of high alarm.

"Yes, I know. They're *my* dogs—that's my boat," I shouted back.

"Oh…well, they've been barking. Do they need food?" she asked.

What? *Do they need food?* Hello, lady. My 30-ton steel boat is floating, uncaptained, across the cove, and you're focusing on whether my dogs are being properly fed?

"No, they have plenty of food. They're fine," I responded through gritted teeth. It hadn't occurred to her that they might be barking because they were drifting across the cove? *What's the matter with people?* My terseness must have communicated itself, because Biff moved his pipe to the other side of his gin-flushed face and began the row back to shore.

As I clambered aboard the *Bossanova*, it became clear to me that no damage had been done. The boat wasn't aground or even up against the shoreline. By the time the harbormaster arrived, I had already hauled up the anchor and motored farther off. He motioned for me to follow him, then pointed to a spot for me to reanchor. I pulled ahead, pointing the bow into the wind. As the boat was pushed back, I dropped the anchor and paid out the line, making sure I felt the familiar tautness that indicated the anchor had grabbed bottom.

Back in the launch with Matt, I realized suddenly that this was the very first time I'd handled my boat without someone else aboard. I had taken the boat out and brought it into the dock every single day on our three-week trek up the coast. But even though I was running everything, there'd been comfort in having John there—another set of hands for the lines, another pair of eyes and ears for the horizon and radio. Deep down, despite our weeks at sea, I had secretly and subconsciously doubted my ability to handle the boat solo. I looked down at my still pristine white shirt. The bottom of the cove was coated with a dark green clay that was sloppy as hell. I'd managed to haul the anchor, move the boat, reset the anchor and head back to dinner in under forty-five minutes, without getting anything on me. Yes, what had started out as an embarrassing incident now definitely felt like a small, albeit embarrassing, personal triumph.

Bouncing over the water on the way back to the town dock, I was happy and relieved. Sooner or later, everyone drags anchor. I had that out of the way now, and I'd been lucky enough to have it happen in a quiet, empty cove. As Matt and I talked, I also figured out what had happened: there was an unusually high tide that day and the extra length of line required to reach the bottom from the (now higher) bow had been just enough to break the anchor from its hold. So that's why it held for two days and then gave up. I knew it was always better to err on the side of too much line rather than too little (unless you're close to other boats and can swing into their paths, of course), and I had sloppily underestimated as I paid it out the last time. I wouldn't do that again.

A few days later, I was at my slip and fondly eyeing the *Bossanova*. She desperately needed a new paint job, and I was feeling falsely flush with the extra money I was making painting with Jay. I talked to Rick, who agreed to haul and block the *Bossanova* so I could paint her in his yard. Boatyards generally do not allow do-it-yourselfers because they prefer to charge you obscene amounts of money and do it for you. I was absolutely thrilled by this act of kindness, because otherwise the *Bossanova* could never have been painted.

I decided to go with a navy blue color for the hull and to paint the rub rail bright orange. The navy seemed classic and had always been my favorite color. I also thought it might have the same slimming effect as a little black dress. The bright orange was jaunty and a nice reference to the internationally recognized "safety orange." Orange would be a nice nod to the *Bossanova*'s salty, working-class lineage.

And so, a few weeks later, *Bossanova* was hauled out in Rick's Travelift, a giant sling on motorized wheels. I had warned him that the boat weighed 30 tons according to the survey, which was the exact weight limit of his lift, but he seemed unfazed. "Yeah," he said. "I bet it weighs less than that if you're not all loaded up with fuel and other stuff." Rick knew his boats, so I believed him.

Now, as they hoisted my baby out of the water, you could cut the tension with a knife. Rick was mopping his brow and cursing under his breath. There were some creaks and groans. The pilothouse roof just barely fit under the forward beam of the lift. But Rick was right—the weight wasn't an issue so much as the width and depth of the vessel. Out of the water, the *Bossanova* was shockingly large—it was easy to see why she was so roomy inside.

The Travelift moved the boat into a spot behind the large

corrugated metal hangar where smaller boats were stored, and the guys propped the boat up with a series of metal jackstands that leaned beneath the waterline of the hull.

The next day, I borrowed my brother Tom's grinder and went to work on a couple of rust spots. The *Bossanova* had some flaking rust around her old stern davits but was otherwise almost rust-free. After I'd ground them down, I treated these spots with sticky rust-inhibitive primer. Then I sanded the entire hull with medium-grit paper using a palm sander. When I was ready to prime, I used WillBond, a liquid-chemical bonding agent, to tack all the sanding dust off and give the surface some extra bite.

I stirred a dark blue tint into the white primer, which resulted in a garish aquamarine hue. Oh, well. I taped the rub rail on the top and bottom and then rolled the entire boat with primer. When I was finished, it looked like some insane Caribbean fisherman's wet dream. More than one member of the peanut gallery stopped by to say I should leave it that color. I thought about it—it was certainly festive looking—but it was just *too* much. The *Shady Lady* could have carried off this party dress, but I imagined the *Bossanova* as classic yet sporty. No loud Hawaiian shirts or tube tops for my beautiful boat.

Instead, I was shooting for ruggedly handsome. A bright sky blue would seem like the raised middle finger of a high school misfit with a dyed blue Mohawk: Hey, look at me: I'm *intentionally* ugly.

I left the boat to dry. Luckily it didn't rain until the following day, when the primer had already had a chance to harden. I took advantage of the weather and busied myself with painting the salon and galley interiors. I replaced the off-white with a slightly warmer putty color, while I listened to the downpour drumming on the steel overhead. It was a wonderful sound—like rain on a car roof but richer, deeper.

After the *Bossanova* had another day to dry in the sun, I went back and rolled the hull with dark blue—this time I skipped the sanding, which just seemed to scratch up the primer coat. Instead, I used more WillBond to reprep the surface before I applied the paint. The following day, I gave it another coat of navy blue and slapped orange paint on the rub rail. Voila! The *Bossanova* had been transformed. The whole project took me about four working days and $800, including the haul and launch. The swell of pride I felt as I watched her being lowered back into the water wasn't even marred by seeing the belts on the lift scraping the fresh orange paint off the rub rail. No big deal. That's what rub rails are for.

All I needed to do now was put the name on the stern properly.

I was having an amazing summer. Happy in my part-time work, enjoying the boat, still savoring the triumph of the big trip. I was on top of the world.

Sometime this week, in the middle of August, I met Lars. I was still painting the boat and I'd quit early to go to a dinner party on the beach.

Even before the sun had set, you could see it was going to be a gorgeous starry night. A big grill was going, covered with sweet corn, lobster, chicken and hamburgers. A bar with an

endless supply of expensive rosé was well-tended and minia-
ture crab cakes with aioli were being circulated by cheerful
college kids with summer catering gigs. A long table was set
for about thirty and a bonfire had just been started. I was
tanned and relaxed, with friends, awed by the beautiful night,
deeply aware of my good fortune and glad to be alive.

"You've got to meet Lars," my hosts had said. They'd just
returned from a day of sailing with him and seemed slightly
giddy. "You guys will love each other, he's been a captain for
years, you will have so much to talk about. And, wow, is he a
chick magnet! But don't worry—we already warned him about
you."

Experience led me to expect a guy who had gracefully
passed middle age—an old salt with a slight beer belly and a
wind-burned complexion whose "chick magneticism" derived
more from the twinkle in his eyes, acres of charm and great
sea stories than from an Adonis-like physical presence. So I
was shocked when someone said, "Mary, this is Lars." I had
already noticed him standing alone at the bar in khakis and
an Icelandic sweater and wondered vaguely who he was. He
was good-looking in a young Bobby Kennedy way.

We chatted through hors d'oeuvres, sat beside each other
and talked all through dinner and then moved to the bonfire
and outlasted the caterers and most of the guests while we
continued the conversation with the last bottle of wine. We
talked about my experience at Chapman, my trip up the coast
and each place I had stopped along the way. He told me about
his years as a first mate on a research vessel that was mapping
the ocean floor, about running a 90-foot boat for a Hungarian

businessman, about overseeing the construction of the friend-ship sloop he was captaining that summer.

I knew he had a girlfriend. He knew I was gay. But he asked if he could come see my boat and we wound up kissing in the pilothouse. We left the clichéd trail of clothes on the way to the stateroom and got very little sleep.

I've been gay my whole life. No question about it. I like men, I have many male friends, I had plenty of romantic op-portunities with men available to me—my heart just felt drawn to women. Living in Manhattan and working in pub-lishing, I'd been able to be myself. I didn't advertise my love life, but I was open about it. Like everyone else, I kept a photo of my beloved on my desk. On Mondays when people talked about their weekends, I talked about mine, too. Consequently, everybody knew I was gay, even most of my acquaintances. And nobody seemed to care, which was as it should be.

So on this particular morning, it was pretty darn odd to wake up with the usual two fur balls wedged against me and this...*man* in the bed. He was naked, sound asleep and crowd-ing me over toward the edge of the mattress. When I looked at him, I felt fond but puzzled. What was this about, I won-dered, grabbing a robe and moving to the forward stateroom for more bed space, and perhaps some mental space, too. I couldn't fall back asleep, so I stared at the ceiling and mulled things over.

When I first realized I was attracted to women, I told myself it was a phase. I was just doing some open-minded experimenting. Complete balderdash, and I knew it even then. When I had accepted that this just seemed to be who I was, I tried explaining to myself (and sometimes others) that I was totally open to the *idea* of men, but so far hadn't met one I could fall for. At some point—though this seemed technically true—the absolute consistency of my behavior made it pretty clear that I was just...gay. And I was comfortable and happy with that. My parents, who were very liberal in every way, were nonetheless dismayed. My father said he worried that life would be harder for me, but I think my mother was just appalled, even though she's always tried to be accepting.

I hate all the misunderstandings about homosexuality. It's not a "choice" any more than heterosexuality is a choice. It's certainly not a "lifestyle." And it's about so much more than "sexuality." It's about romance and intimacy and love. It's about where your heart wants to go, not just your lips and...other parts.

So, I wasn't sure how to explain the seminaked man who now stood in the doorway. "What are you doing in *here*?" he asked, in a slightly hurt tone.

"You were totally hogging the bed, so I came in here for a little space."

"Oh, good!" Lars said. "I was afraid you woke up and thought, "'What am I doing with this fat, ugly man who smells bad?'"

"You don't smell bad," I responded with a smile.

"Ha-ha," he said. "Can I get in there with you for a few minutes before you get up and make us both coffee?"

We lay there and talked for a while. And one thing led to another again. This time, I didn't have the excuses of a little too much rosé, or the romantic spread of stars above, the cozy pop and hiss of a warm bonfire on a cool night to blame. I can't be more explicit about the sex: I'm modest by nature and saying this much already feels like I'm taking all my clothes off in front of you. (Ignore that little roll around my waist—three months of doing nothing but sitting and writing. I'll lose it soon. Really. I will.) Let me just say that my physical relationship with Lars was excellent. We had chemistry. I was attracted to him, so I was attracted to his body. It seemed that simple and natural, despite the lack of precedent.

After coffee, Lars had a boat to move and I had to meet Jay to paint. We were halfway through transforming a long room into a long limestone room. He was a genius at what he did, and I liked being his handmaid of hues, his sous-chef of semigloss.

Of course, I tell Jay everything, and it was fun to have someone to giggle with about Lars. I mean, that was my primary reaction: I thought it was funny that this had happened after twenty-some years. And what was even funnier was that it was...great. But I also absolutely, positively had no expectations or desire for anything else to come of this...interesting accident.

So I was surprised by the sweet message Lars left me that afternoon. "Hi, I'm at Sunset Beach, having a little calamari and rosé. I'd like to play pétanque, but I've got no one to play with. Well, okay. Talk to you later."

He called a few more times without leaving messages, and I felt pleased but also increasingly anxious, unsure of what was happening. At the end of the day, Lars dropped by my high-and-dry boat, which was still propped up in the yard while I finished painting it.

"Boy, are you pushy," I greeted him. "And look at me…" I waved a hand over myself. I had on a paint-spattered tennis shirt and dirty blue track pants, fresh (or not so fresh) from the jobsite. Not pretty.

"You look beautiful," he said.

"Well, it's awfully dark out here," I countered. But honestly, I couldn't believe how handsome he was. His eyes were such an extraordinary blue that they seemed lit up from behind. And he had great features—an aquiline nose that had once been broken, full lips, short brown hair that was dusted with silver. I later saw photos of him in his twenties when he'd been too good-looking, as pretty as a male model. He was 38 now and I liked the way time had roughened his smooth edges. His imperfections added to his appeal.

"I guess I just wanted to tell you that I really hope you don't regret what happened last night."

I told him that I absolutely did not regret it. I was glad it had happened. And that was true.

He continued. "Yesterday, the whole day, everything felt very fateful. I don't know how to explain it…it was just a great day, and then I met you. I hope you want to keep seeing me…and having sex…" he smiled, "but if you just want to be friends, that would be okay, too. I want us to hang out together, at least. I just like you. I think you're very cool and I feel strangely bonded to you"

This was a mouthful I hadn't been expecting. I had no idea what to say. I thought guys weren't emotive. I thought really handsome guys with girlfriends were especially un-likely to "bond" after one night with someone. It had been a fling. Just one of those things, as Cole Porter would say. Hadn't it?

I told him we would definitely be friends. I thought he was a great guy and last night had been amazing. But I had been gay for forty years and he had a girlfriend, so anything else seemed pretty unlikely.

Shows what I know.

CHAPTER NINE

No emotion, any more than a breaking wave,
can long retain its own individual form.
—HENRY WARD BEECHER

A couple of evenings after our chat on the stern of the *Bos-sanova*, Lars and I planned to meet for a friendly dinner. I was trying to manage my vague discomfort with the situation, and keeping it light and friendly seemed the way to go. I was later than I meant to be when I parked at the Coecles Harbor Marina on Shelter Island and walked across the yard, looking for the *Friendship Sloop*. I found its mast first and my eyes followed it down to the deck. And there was Lars—turning the pages of an open book that was spread against his knees, with a glass of rosé in his other hand and his whole profile lit up by the sunset behind him. It was a sight that made me smile involuntarily, and it was right about then that I thought, *Oh my god, maybe this* is *going to be a date.*

After a very quick tour of the small boat, we headed for

the neighborhood restaurant that had become his haunt over the summer.

Sunset Beach is a little piece of St. Barth's transported to the Northeast. A groovy ground-floor bar space faces out at the beach. Above that are several levels of open patio areas, strung with festive paper lanterns and open to a great view of the sound. The staff is young and French. The music is always in a style I think of as international hip—very atmospheric but you'd be hard-pressed to hum a few bars. Heat lamps stand like vigilant sentinels, ready to turn away the chilly night breeze. Sunset Beach is always packed and the prices are obscene. But it is like a trip to the French Caribbean—dollar for dollar, if you don't order dessert—and worth every penny on a moonlit summer night when you want to have fun.

Oh, the fun we had. We had to wait at the bar until our table was ready, but it didn't feel like waiting. We picked up our conversation where we'd left off before, and by the time the host came for us, we'd forgotten all about eating. When I turned to pick my jacket off the stool I'd been sitting on, I found myself planting a light kiss on Lars. I didn't know where that urge had come from, but it just felt right.

"Hello," he said with a grin. "I guess this is going pretty well so far."

It was, I realized suddenly, not only a date but my first real date with a man. I couldn't count that guy in high school who came to our door in a coonskin cap with a shotgun. I know, you think I'm joking, but unfortunately I'm not. This was upstate New York, and he had been out "checking his traps." No wonder I was gay.

So how did I feel about sitting across the table from a funny, good-looking guy on my second date, a mere twenty-five years later? Absolutely normal. No different than if I had been sitting across from a funny, good-looking girl. And that was the only surprising thing about it: it just wasn't surprising at all. What if what I had said all those years ago really was true? What if I had always just happened to be attracted to women, and now I had finally met a man who fit my unique idea of lovable?

It was quizzical but I wasn't worried. What I was a little concerned about was his girlfriend situation. I know this may be hard to believe, coming as it does from a lesbian on a date with a man, but I'm fairly square. I don't like messiness, I don't like overlap. Love is complicated enough without deception. I had never cheated on anyone, and I wasn't sure I wanted to be a party to someone else's cheating either. But as I was soon to learn personally, love makes its compromises.

And so our surprising and aimless affair began. In the weeks that followed, I found some way to excuse myself from worrying about Adele, Lars's girlfriend who lived in Florida. At first I felt a guilty sense of sisterhood betrayed. But I rationalized that whatever deception was going on was between the two of them, and had nothing to do with me.

At the same time, I was no fool—I never fully let my guard down and forgot about Adele's existence. Lars and I were having more fun than seemed legal, but spoken romance was off-limits. It was as though we had agreed that that belonged, rightfully to Adele. Still, it's hard to keep a lid on love. Every now and then, one or the other of us would slip up and let some deeper feeling spill over the edge of our self-containment.

The magic place that Lars and I seemed to inhabit together slowly worked its spell. It was a big world: we talked about books, music, our childhoods, places we'd been, places we wanted to go and, most of all, politics. We were kindred spirits in our liberalism, in our despair about American culture, in our outrage about Bush and the erosion of civil liberties. When we weren't working ourselves into a lather about the state of the world, we were laughing, drinking more than we should have and fooling around. Within a month, we figured we had almost made up for the previous four decades I had spent estranged from male anatomy.

A couple of weeks into all of this, we were at Sunset Beach on a gorgeous Sunday afternoon drinking rosé (again) and soaking in the sunshine.

"I do love you, you know." Lars said suddenly.

"I know you do. And I love you."

He looked me in the eyes and smiled, nodded his head and raised his eyebrows.

It was a sweet moment, as natural, relaxed and easy as the warm sun on our faces.

Hey, maybe I'd been gay my whole life but even I knew that guys don't generally like to admit this kind of thing. Then again, Lars was not your typical guy. He was a real man's man—there was nothing metrosexual about him. But he also genuinely loved strong, independent, interesting women. He loved sex but had not even a tiny, venal interest in the Paris Hiltons or Pamela Andersons of the world. He hated professional sports, with the exception of cycling (his passion in life) and occasional tennis. Boats had long ago become more of a livelihood than a love, though he was very accomplished and took pride in what he did. Lars liked pornography, but he also liked going to the opera. He loved both the Grateful Dead and waltz music. He liked five-star hotels and camping. He would push the limits of his body with triathlons and long road races, but he also liked, occasionally, to sit by the pool and drink all day. Lars was truly a free spirit, but I also sensed that his lifestyle sometimes made him feel lonely, that he needed connection to someone to give him a sense of place.

We had these (and many other) qualities in common, though probably in reverse proportions. I suspected that Lars believed that adventure was what happened to you alone, and that love, though he believed in it and needed it, was the natural enemy of adventure. I loved a life of adventure, but deep down, I still believed that life with the right person could be twice the thrill, the greatest adventure of all.

I had held to this stubborn belief like a cactus in the desert, despite my bad luck in love. It wasn't that I wasn't capable of happiness alone, because I'd certainly been alone, often and happily. But my parents' marriage had poisoned me with the belief that true love was what it was really all about.

After more than forty years, my parents still left little notes for each other around the house. One year my mother gathered some up and laminated them, then cut them into shapes that fit into the soles of my father's fishing boots, so he could take the notes with him without getting them wet. Another time, my mother was regaling us (as she is wont to do in excruciating detail—I once teased her that she was the reason I became an editor) about a dream she'd had. In the dream, she was wearing her grandmother's pearls when the string broke and they spilled all across the floor. My very down-to-earth father noticeably blanched. "What?" my mother asked him. "I dreamed I was on my hands and knees picking up pearls," he said.

I know—it's sickening. What's worse is that they were married while they were still in college. My earliest memories are of my father doing handstands on a skateboard in our Philadelphia apartment.

Of course, their relationship isn't perfect. I know they've had their fair share of ups and downs. The truth is, they're totally codependent, but it works for them, so who cares? For years, I wished for a relationship that was half as happily dysfunctional as theirs.

It took me a long time to see an accidental downside to my parents' lifelong love affair. It had created a small but intense emotional vacuum in the rest of the household. As much as they loved us kids and put a lot of thought and energy into educating us, there was a way in which my parents' love for each other felt unintentionally exclusionary. We were always on the outside, watching through the window as they danced to music we couldn't hear. Home was a great place, but we were subjects of that kingdom, not citizens. We moved so often—every two years, at least, throughout my entire childhood—that home for me was less about a place than about these loving and slightly eccentric people with whom I lived.

As a result, when I first left my family to go to Brazil (and for many years after), I felt lonely and deeply compelled to create my own family in some form.

Consequently, I've led a life of serial monogamy, otherwise known as one failed relationship after another, though—you know me by now! I prefer to think of it as a series of short successes. I wish the love of my life and I had met at 20 and stayed together. But it's not what fate had in store for me. I've looked long and hard at myself to discern a pattern, but I can't find one. To paraphrase Tolstoy, happy relationships are all alike: every unhappy relationship is unhappy in its own way.

This thing with Lars puzzled me at times. It was different, no doubt about it. It was comfortable, less intense—as though the innate differences between a man and a woman wedged a firm pillow between us. There was no merging going on, no blurring of identities that so often happens between women. I liked it. We were very similar in lots of ways, and we gave each other plenty of space. Most of all, it was fun.

As summer drew to a close, Lars talked of going to Maine to shingle his barn. We had already agreed that he would help me with the last, short leg between Sag Harbor and Maine that I needed to make aboard the *Bossanova*. Now, though, I looked beyond those next few months. I needed a plan. Where should I go next? The Bahamas seemed appealing, but as I already knew, it was a long journey and I would need crew. So I made Lars a proposition: What if I stayed and worked with him for several weeks on his barn in return for his help taking the *Bossanova* south? It was a deal that was beneficial to us both, and he quickly agreed.

Lars went to Maine a few weeks before me to get started on some projects. One day as we talked on the phone he said, "You know, I don't want you to think this is just a business deal—that it's all about you coming up here to be my barn slave. Yeah, that'll be great, but I've been thinking about how much I want you to meet friends I've had a long time, meet my mother, hang out at my house. You know what I mean?"

Of course I did.

WE SAID GOOD-BYE TO Sag Harbor on a Wednesday afternoon and hello to midcoast Maine about thirty-two hours later. The weather was perfect and we timed the trip to have the current with us. We took turns standing watch, passed through the Cape Cod Canal and ran all night. At around 1600 hours we arrived and tied the *Bossanova* up at a local marina that had already emptied for the fall.

The first place we went was Lars's house. The Block Cottage, as it is known, is perched dramatically on a shelf of rocks, with almost 360-degree water views. It's an old-fashioned shingled summer cottage. One year when he was away at school, Lars's grandmother had misguidedly renovated the downstairs, to his everlasting chagrin. Except for the stunning ocean views from every window, and an old fashioned, homey kitchen that was left untouched, the downstairs had been dry-walled and redone in a modestly formal way that would have been at home in any good suburb. On the other hand, that's like saying except for the big nose, Jimmy Durante was handsome. The views were staggering. Upstairs, the house

retained its original charm and was wonderfully unpretentious: painted floorboards and bare wood walls, wallpaper
with sailboats printed in primary colors and an old claw-foot
tub in the dormer-style bathroom. This was the rustic dream
of summertime that Ralph Lauren painted with gloss and
sold to people who didn't know that imperfections were signs
of taste that money can't buy.

Lars's mother rented the house out in the summers and
had built a sweet apartment above the barn where she could
perch. In the winter, she went to Camden, but for the first few
weeks of our stay, she was in the apartment each night. She
was a terrific cook and spoiled us rotten. We gathered in the
Block Cottage kitchen each evening after a full day of hard
work. Lars and I were clearing out his barn, then building a
floor and, finally, shingling; Lars's mother was involved in the
painstaking labor of varnishing a boat. We'd listen to NPR,
drink wine, cook, debate and sometimes even dance. In the
weeks leading up to the election, we also ganged up on Lars's
mother in a relentless effort to convince her to change her
vote. She was a good sport about it, but living with Lars, you'd
have to have developed excellent sportsmanship skills from
way back.

Lars was very funny and he loved to tease. I grew up in a
family where teasing was a comfortable way of expressing
love, but I noticed that out in the larger world, many people
were offended by teasing or sarcasm. Lars and I spoke the
same language.

I remember one day when we were shingling, I was in a
rare bad mood, quieter than usual.

"What's the matter with you today?" Lars asked. "You're not your usual fun self." I told him it was nothing, I was just premenstrual.

He dropped his hammer dramatically. "Uh-oh. You know what we have to do, don't you? We're going to have to use the old Indian remedy. They'd send a woman out into the woods alone for a few days and tell her not to come back until it was over. Do I have to do that, too? Because I will. I don't want any mopeyness around me. I'll just send you out into the woods like a squaw."

He wasn't always funny, but he always made me laugh. Freud be damned, one day I realized that Lars reminded me a little of my father. They had the same sadistic but hilarious sense of humor—they were both tough but sensitive and intelligent. And all of the physical work we did each day no doubt reminded me of my youth. I loved to help Dad with chores: splitting wood, painting a barn, dragging things to the dump—anything I could do alongside him was fun. Now, being outdoors each day as the Maine fall trotted toward winter, I realized that it had been a long time since I'd enjoyed such hard work. It was exhausting but satisfying, and Lars was always good company.

One evening, he stopped by my boat for a drink before we joined his mother for dinner. We were chatting about nothing, listening to jazz—I think I was cutting limes. Completely out of the blue, he rattled the ice in his glass and said, "You know, I'm in love with you."

And I realized it was true. I was no dummy. We wouldn't be here in Maine together for any other reason. We'd had a perfect way to say our good-byes in Sag Harbor if things had really run their course.

This would have been a good time to try and get to the bottom of what all this meant regarding Adele. But I didn't know what to say, and I felt, as I always did in our affair, the sensation of going down a steep hill that should have been wildly exhilarating but wasn't because the brakes were on. It was both wonderful and a shame that it was so much less than it could have been. It was baffling, like a connect-the-dots picture before the lines are drawn. I sensed what it might be but wasn't.

Toward the end of our six weeks in Maine, we spent a very decadent Sunday at the Block Cottage. The gorgeous foliage that had waved to us on arrival was now halfway to mulch, and the bare tree limbs grasped for the sky like arthritic fingers. The house was swaddled in a gray woolly fog. It was raining intermittently, and the waves crashed rhythmically on the rocks below the house. We read the *New York Times* on the living room floor before a fire that snapped and hissed. We shot trap off the front porch and drank red wine and cooked a feast of some sort. It was a perfect day. That night, we were still stretched out on the living room floor talking as the fire burned itself out. Lars was telling me about how he was a little bit tired of his profession, and I was quizzing him about what else he'd enjoy doing.

"Think about it," I said, in career counselor mode and looking for ideas. "Look back. When was the happiest time of your life, and what were you doing then?"

"This summer, when I met you," he answered without hesitation. If I hadn't been lying down, I would have fallen over. And obviously, I had no idea what career that suited him for.

There is one thing I learned a long time ago: there is no point trying to see into the inscrutable heart. Once in a while, Lars and I let ourselves talk vaguely about the future, but I knew enough never to imagine it would actually happen. Lars's plans, like my own, were constantly changing. And if he couldn't go to Miami for a weekend without feeling torn about Adele, he wasn't going to make it through a few months.

Every time I went over to the Block Cottage to check my e-mail on Lars's computer, I was vividly reminded of my temporary status. The drop-down menu on his AOL sign-in page had his screen name, Adele's screen name, and a third name that was "Guest." That was me. I was Guest, and I never forgot it. Every time I signed in, I felt a little stab of sorrow for the sight but also gratitude for the reality check.

A few days after our perfect Sunday, we put his Land Cruiser away for the winter and took the sails off his boat. I had reluctantly decided that I was too broke to go to the Bahamas after all. We brought the *Bossanova* around to a mooring on the other side of the island and left her to be hauled out, the last boat into the yard for the winter. It was gray and bitterly cold as Lars rowed us toward shore. I watched the *Bossanova* fade out of sight with a lump in my throat. I felt as though I was abandoning my best friend, but it gave Lars another chance to torture me. Stuck mostly indoors for the last few days as late fall dramatically became early winter, Lars took every opportunity to peek out the window at the lowering sky and howling winds. "Aw," he'd say. "Poor *Bossanova*, out there all alone, in the cold sea, by herself, with no one to take care of her or love her."

The first snowstorm of the season came through within days, and 5 inches of white transformed the world. It was officially winter. And though it didn't feel like we were anywhere near over, it was time for both of us to move on.

CHAPTER TEN

There exists only the present instant...
Now which always and without end is itself new.
There is no yesterday nor any tomorrow,
but only Now, as it was a thousand years ago
and as it will be a thousand years hence.

—MEISTER ECKHART

I know, you can hardly stand the suspense. Does the book have a fairy-tale ending? Does true love conquer all?

I'm happy to report: yes and yes.

But Lars and I did not sail into the sunset together. For six weeks or so after Lars's return to Miami, we remained pretty attached to each other—frequently e-mailing and calling. Once in December, in the middle of the night, the phone rang, and when I answered, I heard just music, our song. I knew he still loved me then, and the happiness that short message caused made it clear I loved him, too.

But by January, things were changing. We kept in touch,

but mostly by e-mail. He was busy with a celestial navigation course. I was engrossed in a writing project. In February, his mother casually mentioned in an e-mail that Adele was going to quit her job to join Lars in his new position as captain of a 97-foot sailboat bound for Europe.

Lars called me soon after that, when he had just passed a very tough Coast Guard exam for a license he'd inadvertently let lapse that fall. ("It was your fault," he accused me. "You distracted me with your shingling.") He had been drinking rosé and sounded happy but nostalgic.

That's exactly how I feel whenever I think of Lars and our time together. The huge gift of this relationship was its perfect impossibility, its planned obsolescence. My affair with Lars was just me seeing my very last, cherished, concrete notion of self go out the porthole.

However, before my mother gets too excited by heterosexual possibilities, let me be clear that I am still gay. Sorry, Mom. The fairy-tale ending is that I have finally met my match: she is beautiful, sweet, smart, talented and crazy about me. I believe we will grow old together and that I will never shed another tear of loneliness or self-pity when I see a sappy toothpaste ad. Now that I have found her, she seems as inevitable and gorgeous as the sunrise. Everything is as it was meant to be. But it always was.

There were really no missteps on my road to here. I derived something important from all my relationships, though at the time all I wanted was permanence. Meister Eckhart wrote "When you are thwarted, it is your own attitude that is out of order."

Sure, you're thinking—easy for me to be philosophical about it now that I have found love. But I found love *after* I realized these things, after I felt a sense of peace and rightness with *now*.

Example: About a year ago, I was very worried about money. I was parked outside the Seafood Shop in Wainscott, taking a lunch break from some painting I was doing for a friend. I was sitting in the car I couldn't afford anymore with the top down, getting those first wan but delicious rays of spring sunshine on my face and eating a delicious $4 tuna roll.

I had spent the morning obsessing about buying a house and realizing how impossible that was now. Not enough income, not very good credit, and so on. I felt myself becoming anxious and depressed about the future. And then I pulled myself back to my very humble present in the sunshine and felt lucky. Okay, I'm poor. But I won't be forever. It's a beautiful day and I'm not sitting in a conference room in a suit listening to the deluded opine earnestly about things that don't matter. I'm where I want to be, doing something I enjoy. Right this second is wonderful.

A sense of meaningless I have felt at different moments throughout my life, the issues I have always had with loss, were a big, albeit subtle, part of what sent me off to sea. I'm intensely sentimental, insanely loyal. In the past, there has always been something retarded about my grieving process. I can't let go. I don't move on emotionally as easily as other people seem to. It's caused me a ridiculous amount of woe, but I haven't known how to change that about myself. In fact, painful as it was, I wasn't sure I wanted to change it. I've never understood people who move seamlessly from one relationship to the next without ever being alone or hearing themselves think.

Yet I was able to let Lars go much more easily than I have most other relationships. I don't think that's because he is a man, and I know it's not because I loved him any less. But when our relationship began, I knew it would probably not last forever. And so, when it ended, I was sad and I missed him, but I did not feel the same acute sense of failure that usually haunts me.

This was a revelation. I finally understood that my tendency to engage in lingering grief was much more about mourning the end of a dream than it was the loss of an actual relationship, which hadn't worked out, after all. Why did I freight each romance with the need to be eternal? Why couldn't I just see each of them for what they were—successful within the limits of their potential?

You may have figured this out years ago, but not me. I had no idea. Maybe this perspective was the dark side of my über optimism. Perhaps I had always expected too much, but let's face it: we are trained to. Our culture inundates us with images of the happily married and reveres the notion of everlasting love, despite the divorce statistics. Maybe we'd all be happier with more realistic expectations.

I recently came across a Bertrand Russell quote that sums up one of my deepest beliefs: "The past is an Awful God, though he gives life to almost the whole of its haunting beauty."

This past year has changed me, though, and I've been pondering another thought of André Gide's: "Through loyalty to the past, our mind refuses to realize that tomorrow's joy is possible only if today's makes way for it; that each wave owes the beauty of its line only to the withdrawal of the preceding one."

I mention these because separately they are lovely quotes and both quite true. But together, they are the one-two combination to the mystery of my life, the key to an understanding that has brought me a lasting sense of happiness.

An unforgettable moment happened when I was 5 years old; I wish it had been more immediately formative than it was. My grandfather took my younger brother, my cousins and me on a walk to the barnyard. As we gazed in at the sheep milling around the base of the silo, he said "I will give a marshmallow to the first person to grab that hotline." My brother and cousins immediately laughed and said, "No way! We'll get a shock! You can't trick us." But me, I was just old enough and mature enough to overthink it. "Hey, this is my grandfather. He wouldn't tell me to grab an electrified wire if it was really on, would he?" So I did. And it was.

Maybe my whole life has been a little like the hotline incident. Maybe things *are* as simple as they seem. I just suffer from wanting everything to make more sense than it does. Now I see that my life has plenty of meaning, even though it will probably never make sense.

Don't get me wrong: I'm not claiming to have arrived at some enlightened point where everything is clear to me, where I have it all figured out. Far from it. I am, however, often reminded of a favorite bit of knowledge I gleaned in seamanship school. All my life I'd been walking along beaches, watching the surf wash up on the sand, thinking about the journey that water made to finally lap thirstily against the dry shore. But a wave is not a moving mass of seawater. It's just still, stationary ocean with energy moving through it. I now see time as that same illusory wavelike movement through the present.

There was an ordinary moment that I often recall, walking down Christopher Street with Leslie. We were both on our way to work, and she reached out and held my hand for an instant. I can't tell you why, but a regular day became sublime—that simple touch of love filled me with a sense of contentment, of rightness, that haunted me for years. For years, every walk down every other street became a walk without her hand in mine.

Somewhere that is still alive in my heart, my grandfather and I stand before a sweltering barbecue pulled out onto the gravel in front of the barn. Crickets are singing, dusk is falling, the bats and swallows have left the rafters in search of evening. Ham has on khaki shorts, a broad worn leather belt, a chambray shirt and suede chukka boots. He spits a long stream of clear rum into the kettle that causes a burst of thrilling fire. God, I love him.

It's another time, and I am 5, tucked in my summer bed, the faded rose wallpaper all golden in the late setting sun. My father has come up to say good night and on his way out, he accidentally walks into the door, trips over the rug, pretends to fall down the stairs, and each time he returns (about half as old as I am now!) to say sternly, "Stop laughing. Get to sleep. I'm serious," before he does it one more time. Almost forty years later, the memory of this little act of love for me still makes me smile.

And the Sunday afternoon that Lars and I spent drinking rosé, when his eyes out-blued the cloudless sky and he told me he loved me, that's not gone. It's just not happening now.

There's nothing to mourn. There are wonderful things ahead of us as surely as there are sad ones. What's important is to be alive, to feel these things deeply as they happen because they make us—we are just blood, bone, guts and the sum of these moments. They become who we are: individuals who are mysteriously flavored by our past, who live every minute with the consequences of who we have loved and who has loved us.

And the awful pain left by those who didn't love us well enough or long enough? Well, that falls away, doesn't it? It withers like an immaterial husk, and always reveals something else: a necessary lesson, a lifelong friendship, an unforgettable few months.

I owe my peace to the *Bossanova*—a small steel ship, but also a magic carpet. The joy I felt standing on her bow I will always be able to summon at will, as though I am still right there with the breeze in my hair and the sun on my face. I know now that this moment—like every moment—is within me: still real, very alive, the magnificent now that is also the invisible doorway to whatever comes next.

ACKNOWLEDGMENTS

It's not hard to know where to start, or where to end, but I'm sure that I'll fail to thank someone who should fall squarely in the middle. So, to you (and you know who you are) as well as to everyone I met along the way that shared in my adventure, thank you.

John McIninery, first mate extraordinaire and a heck of a human being, words (which I think very highly of) just won't suffice. I couldn't have done it without you. Call me, you knucklehead!

To Collin Janse van Rensburg for his generous tutoring and crewing, Carol Gordon for her expert insurance help and infectious *joie de vivre* and all my Chapman classmates for their camaraderie, thank you so much! Susan Scheer provided friendship and support above and beyond the call of duty, and I can never thank her enough. Captain Bob Swindell, thank you for giving me the confidence to leave the dock and the competence to return. Thanks also to everyone at Hinckley in Stuart for being such friendly hosts, and to everyone at Ship-Ashore in Sag Harbor. Very special thanks to Jeff Poole for his generous spirit and a chivalry that not even lousy weather could dampen. Melville Traber built a magnificent boat that I worked hard to be worthy of, and I'm grateful for his trust.

Henry Alford, Terri Costello, Holley Bishop, Leslie Lubchansky, Erika Mansourian, Laurie Mezzalingua, Pat Bates, Isabel Botelho Leal, Vicky Homan, and Frank and Barbara Sain have each provided great friendship and encouragement. Thanks to Eric Lemonides for pilot services and for making me feel completely at home when I came for a week and stayed for six.

Special thanks to Leslie Klotz and the guys for their absurd, yet deeply touching, belief in my abilities. It has meant the world to me. Chris Peterson, thanks for being there, day in and day out. You've been a lifeline. To my beloved Famiglia Grau-Stern, wooooooo-hooooooo! And to Jay "Jay-Lo" Lohmann, thanks for being the sweetest and making even the most ordinary days fun.

Michele Christensen deserves multiple medals—for her delicious cooking, her warm hospitality, her overall generosity and for being the kind of friend that becomes family. In other words, you're stuck with me. I hope.

Thanks to David Hirshey for believing in the book and to Nick Trautwein and Kate Hamill for their hard work and helpful insights. Thanks to Gail Ross for having my back and to Kara Baskin for her enthusiasm and perspective.

Heck and Samba can't read (yet), but I still feel the need to publicly thank them for joining me on my adventures. They are the world's most adorable deckhands in every kind of weather. Massive love and gratitude, also, to the mighty *Bossanova*, for safely taking me on the greatest adventure of my life. To Mom, Dad, Paddy and Tom, thank you for everything.

Hosannas to the amazingly lovely and talented Karyn Olivier for being my home.

And most of all, thank you, thank you, thank you to Hamilton South whose astounding generosity and love have enriched my life in a million ways.